Official Journal of the International Pharmaco-EEG Society (IPEG)

Neuropsychobiology

International Journal of Experimental and Clinical Research in Biological Psychiatry, Pharmacopsychiatry, Biological Psychology/Pharmacopsychology and Pharmacoelectroencephalography

Founded 1975 by J. Mendlewicz (Brussels).
Since 1983 integrating 'International Pharmacopsychiatry',
founded 1968 by F.A. Freyhan (New York), N. Petrilowitsch (Mainz), P. Pichot (Paris)
Section Editor Pharmacopsychiatry and Associate Editor 1990–2006 B. Saletu (Vienna)
Founder, Section Editor Biological Psychiatry and Associate Editor 1975–2007 J. Mendlewicz (Brussels)

KARGER

Printed in Switzerland
on acid-free and non-aging
paper (ISO 9706) by
Reinhardt Druck, Basel

Appears 6-weekly:
2 volumes per year
(8 issues)

Neuropsychobiology

Guidelines for Authors

Submission

Only papers written in English are considered and should be submitted using the online submission website at: www.karger.com/nps. Should you experience any problems with your submission, please contact

Prof. W. Strik
University Hospital of Psychiatry
Waldau
CH–3000 Bern 60 (Switzerland)
E-Mail: neuropsychobiology@puk.unibe.ch

Conditions

All manuscripts are subject to editorial review. Manuscripts are received with the explicit understanding that they are not under simultaneous consideration by any other publication. Submission of an article for publication implies transfer of the copyright from the author to the publisher upon acceptance. Accepted papers become the permanent property of 'Neuropsychobiology' and may not be reproduced by any means, in whole or in part, without the written consent of the publisher. It is the author's responsibility to obtain permission to reproduce illustrations, tables, etc. from other publications.

Arrangement

Title page: The first page of each paper should indicate the title, the authors' names, the institute where the work was conducted, and a short title for use as running head.
NB: Authors wishing to preserve the phonetic meaning of diacritics (PubMed reduces diacritics to their root characters) must spell their names accordingly when submitting manuscripts (e.g. Müller should be Mueller).

Full address: The exact postal address of the corresponding author complete with postal code must be given at the bottom of the title page. Please also supply phone and fax numbers, as well as e-mail address.

Key words: For indexing purposes, a list of 3–10 key words in English is essential.

Abstract: Each paper needs an abstract in 4 paragraphs and not exceeding 250 words.

Footnotes: Avoid footnotes.

Tables and illustrations: Tables and illustrations (both numbered in Arabic numerals) should be prepared on separate pages. Tables require a heading and figures a legend, also prepared on a separate page. For the reproduction of illustrations, only good drawings and original photographs can be accepted; negatives or photocopies cannot be used. Due to technical reasons, figures with a screen background should not be submitted. When possible, group several illustrations in one block for reproduction (max. size 180 × 223 mm) or provide crop marks. Electronically submitted b/w half-tone and color illustrations must have a final resolution of 300 dpi after scaling, line drawings one of 800–1,200 dpi.

Color Illustrations

Online edition: Color illustrations are reproduced free of charge. In the print version, the illustrations are reproduced in black and white. Please avoid referring to the colors in the text and figure legends.
Print edition: Up to 6 color illustrations per page can be integrated within the text at CHF 800.– per page.

References

In the text identify references by Arabic numerals [in square brackets]. Material submitted for publication but not yet accepted should be noted as 'unpublished data' and not be included in the reference list. The list of references should include only those publications which are cited in the text. Do not alphabetize; number references in the order in which they are first mentioned in the text. The surnames of the authors followed by initials should be given. There should be no punctuation other than a comma to separate the authors. Preferably, please cite all authors. Abbreviate journal names according to the Index Medicus system. (Also see International Committee of Medical Journal Editors: Uniform requirements for manuscripts submitted to biomedical journals. N Engl J Med 1997;336:309–315. www.icmje.org).

Examples
(a) Papers published in periodicals: Chatel J-M, Bernard H, Orson FM: Isolation and characterization of two complete Ara h 2 isoforms cDNA. Int Arch Allergy Immunol 2003;131:14–18.
(b) Papers published only with DOI numbers: Theoharides TC, Boucher W, Spear K: Serum interleukin-6 reflects disease severity and osteoporosis in mastocytosis patients. Int Arch Allergy Immunol DOI: 10.1159/000063858.
(c) Monographs: Matthews DE, Farewell VT: Using and Understanding Medical Statistics, ed 3, revised. Basel, Karger, 1996.
(d) Edited books: DuBois RN: Cyclooxygenase-2 and colorectal cancer; in Dannenberg AJ, Dubois RN (eds): COX-2. Prog Exp Tum Res. Basel, Karger, 2003, vol 37, pp 124–137.

Reference Management Software: Use of EndNote is recommended for easy management and formatting of citations and reference lists.

Digital Object Identifier (DOI)

S. Karger Publishers supports DOIs as unique identifiers for articles. A DOI number will be printed on the title page of each article. DOIs can be useful in the future for identifying and citing articles published online without volume or issue information. More information can be found at www.doi.org.

Author's Choice™

Karger's Author's Choice service broadens the reach of your article and gives all users worldwide free and full access for reading, downloading and printing at www.karger.com. The option is available for a one-time fee of CHF 3000.–, which is a permissible cost in grant allocation. More information can be found at www.karger.com/authors_choice.

NIH-Funded Research

The U.S. National Institutes of Health (NIH) mandates under the NIH Public Access Policy that final, peer-reviewed manuscripts appear in its digital database within 12 months of the official publication date. As a service to authors, Karger submits the final version of your article on your behalf to PubMed Central. For those selecting our premium Author's Choice service, we will send your article immediately upon publishing, accelerating the accessibility of your work without the usual embargo. More details on NIH's Public Access Policy is available at http://publicaccess.nih.gov/FAQ.htm#a1

Self-Archiving

Karger permits authors to archive their pre-prints (i.e. pre-refereeing) or post-prints (i.e. final draft post-refereeing) on their personal or institution's servers, provided the following conditions are met: Articles may not be used for commercial purposes, must be linked to the publisher's version, and must acknowledge the publisher's copyright. Authors selecting Karger's Author's Choice feature, however, are also permitted to archive the final, published version of their article, which includes copyediting and design improvements as well as citation links.

Page Charges

There are no page charges for papers of 5 or fewer printed pages (including tables, illustrations and references). Each additional complete or partial page is charged to the author at CHF 325.–. The allotted size of a paper is equal to approx. 15 manuscript pages (including tables, illustrations and references).

Proofs

Unless indicated otherwise, proofs are sent to the first-named author and should be returned with the least possible delay. Alterations made in proofs, other than the correction of printer's errors, are charged to the author.

Reprints

Order forms and a price list are sent with the proofs. Orders submitted after the issue is printed are subject to considerably higher prices.

KARGER
Fax +41 61 306 12 34
E-Mail karger@karger.ch
www.karger.com

The Guidelines for Authors are available at:
www.karger.com/nps_Guidelines

Neuropsychobiology

General Information

ISSN Print Edition: 0302–282X
ISSN Online Edition: 1423–0224

Journal Homepage: www.karger.com/nps

Publication Data: 'Neuropsychobiology' is published 8 times a year. Volumes 61 and 62, each with 4 issues, appear in 2010.

Subscription Rates: Subscriptions run for a full calendar year. Prices are given per year.
Personal subscription:

Print or Online	**Print+Online combined**
CHF 813.40	CHF 903.40
EUR 581.–	EUR 645.–
USD 737.80	USD 823.80

postage and handling (added to print and print+online)
CHF 50.40 Europe, CHF 75.60 Overseas
EUR 36.–
USD 71.60

Institutional subscription:

Print or Online	**Print+Online combined**
CHF 2324.–	CHF 2556.–
EUR 1660.–	EUR 1826.–
USD 2108.00	USD 2320.00

postage and handling (added to print and print+online)
CHF 63.20 Europe, CHF 94.40 Overseas
EUR 44.80
USD 89.60

Airmail surcharge: CHF 64.– / USD 60.80
Discount subscription prices:
- International Pharmaco-EEG Society (IPEG)
- American Psychiatric Association (APA)
- Association of European Psychiatrists (AEP)
- International Society of Biological Psychiatry

Back Volumes and Single Issues: Information on availability and prices of single print issues and print or electronic back volumes can be obtained from Customer Service at *service@karger.ch.*

Bibliographic Indices: This journal is regularly listed in bibliographic services, including *Current Contents®* and PubMed/MEDLINE.

Photocopying: This journal has been registered with the Copyright Clearance Center (CCC), as indicated by the code appearing on the first page of each article. For readers in the US, this code signals consent for copying of articles for personal or internal use, or for the personal or internal use of specific clients, provided that the stated fee is paid per copy directly to
Copyright Clearance Center Inc.
222 Rosewood Drive
Danvers, MA 01923 (USA)
A copy of the first page of the article must accompany payment. Consent does not extend to copying for general distribution, for promotion, for creating new works, or for resale. In these cases, specific written permission must be obtained from the copyright owner,
S. Karger AG, P.O. Box
CH–4009 Basel (Switzerland).

Subscription Orders:
Orders can be placed at agencies, bookstores, directly with the Publisher or further Karger offices or representatives:

S. Karger AG
Medical and Scientific Publishers
P.O. Box
CH–4009 Basel
Switzerland
(for courier services only:
Allschwilerstrasse 10
CH–4055 Basel)
Tel. +41 61 306 11 11
Fax +41 61 306 12 34
E-Mail karger@karger.ch
www.karger.com

France:
Librairie Medi-Sciences Sarl
36, bd de Latour-Maubourg
75007 Paris
France
Tél. +33 1 45 51 42 58
Fax +33 1 45 56 07 80
E-Mail librairie@medi-sciences.fr
www.medi-sciences.fr

Germany:
S. Karger GmbH
Postfach
79095 Freiburg
Deutschland
(Hausadresse: Wilhelmstrasse 20 A
79098 Freiburg)
Tel. +49 761 45 20 70
Fax +49 761 45 20 714
E-Post information@karger.de
www.karger.de

India, Bangladesh, Sri Lanka:
Panther Publishers Private Ltd.
33, First Main
Koramangala First Block
Bangalore 560 034
India
Tel. +91 80 25505 836
Tel. +91 80 25505 837
Fax +91 80 25505 981
E-Mail panther_publishers@vsnl.com
www.pantherpublishers.com

Japan:
Karger Japan, Inc.
Yushima S Bld. 3F
4-2-3, Yushima, Bunkyo-ku
Tokyo 113-0034
Japan
Tel. +81 3 3815 1800
Fax +81 3 3815 1802
E-Mail publisher@karger.jp

China, Taiwan and Malaysia:
Karger China
Suite 409, Apollo Building
1440 Central Yan An Road
Shanghai 200040
China
Tel. +86 21 6133 1861
Fax +86 21 6133 1862
E-Mail karger.ray@gmail.com

South America and Central America:
Cranbury International LLC
7 Clarendon Ave., Suite 2
Montpelier, VT 05602
USA
Tel. +1 802 223 6565
Fax +1 802 223 6824
E-Mail
eatkin@cranburyinternational.com
www.cranburyinternational.com

United Kingdom, Ireland:
S. Karger AG
c/o London Liaison Office
4 Rickett Street
London SW6 1RU
United Kingdom
Tel. +44 20 7386 0500
Fax +44 20 7610 3337
E-Mail uk@karger.ch

USA:
S. Karger Publishers, Inc.
26 West Avon Road
P.O. Box 529
Unionville, CT 06085
USA
Toll free: +1 800 828 5479
Tel. +1 860 675 7834
Fax +1 860 675 7302
E-Mail karger@snet.net

Change of Address:
Both old and new address should be sent to the subscription source.

KARGER

Fax +41 61 306 12 34
E-Mail karger@karger.ch
www.karger.com

The Journal Home Page is available at:
www.karger.com/nps

A contemporary overview of the most relevant research and treatments in ADHD

Attention-Deficit Hyperactivity Disorder (ADHD) in Adults

Editors
Wolfgang Retz
Rachel G. Klein

Psychiatry; Neurology, Genetics, Psychotherapy, Sociology, Pharmacology, Neurobiology

Worldwide longitudinal studies performed since the 1970s have clearly shown that ADHD persists into adulthood. These findings have stimulated researchers to develop the therapeutic approaches for adult patients, especially in European countries where scientific and clinical interest in ADHD has increased. In this volume, leading experts from Europe and the United States present their long-term results in order to provide an overview of important aspects of ADHD across the lifespan. These results include epidemiology, neurobiology, psychopathology, longitudinal course, comorbidity and social impairment associated with ADHD. Topics include diagnostic problems and therapeutic options as well as molecular genetic studies. Further, morphological and functional imaging studies in adult ADHD are reviewed, as well as the very important issue of comorbidity.
Providing an excellent source of up-to-date information, this publication is essential reading for psychiatrists, neurologists, geneticists, psychotherapists, physicians and other therapists working with ADHD patients.

Contents

www.karger.com/kismh

KI 09464

Key Issues in Mental Health, Vol. 176
Series Editors: Riecher-Rössler, A. (Basel); Steiner, M. (Hamilton, Ont.)
ISSN 1662–4874 / e-ISSN 1662–4882

Attention-Deficit Hyperactivity Disorder (ADHD) in Adults
Editors: Retz, W. (Homburg); Klein, R.G. (New York, N.Y.)
VIII + 192 p., 9 fig., 17 tab., hard cover, 2010
CHF 69.– / EUR 49.50 / USD 69.00
Prices subject to change
EUR price for Germany, USD price for USA only
ISBN 978-3-8055-9237-6
e-ISBN 978-3-8055-9238-3

KARGER

Order Form

Please send: ____ copy/ies

Postage and handling free with prepayment

Payment:
Please charge to my credit card
☐ American Express ☐ Diners ☐ Eurocard
☐ MasterCard ☐ Visa

Card No.: ______________

Exp. date: ______________

CVV/CVC ______________
(3 digits in the signature field on the back of Visa and MasterCard)

☐ Check enclosed ☐ Please bill me

Orders may be placed with any bookshop, subscription agency, directly with the publisher or through a Karger distributor.

Fax: +41 61 306 12 34

S. Karger AG, P.O. Box, CH–4009 Basel (Switzerland)
E-Mail orders@karger.ch, **www.karger.com**

Name/Address: ______________

Date: ______________

Signature: ______________

Neuropsychobiology

More than 60 Years of Lithium in Contemporary Psychiatry

Editor
Janusz K. Rybakowski, Poznan

4 figures, 1 in color, and 8 tables, 2010

KARGER Basel · Freiburg · Paris · London · New York · Bangalore · Bangkok · Shanghai · Singapore · Tokyo · Sydney

S. Karger
Medical and Scientific Publishers
Basel • Freiburg • Paris • London •
New York • Bangalore • Bangkok •
Shanghai • Singapore • Tokyo • Sydney

Printed in Switzerland
on acid-free and non-aging paper (ISO 9706) by
Reinhardt Druck, Basel
ISBN 978–3–8055–9410–3
e-ISBN 978–3–8055–9411–0

KARGER

Fax +41 61 306 12 34
E-Mail karger@karger.ch
www.karger.com

Neuropsychobiology

Vol. 62, No. 1, 2010

Contents

KARGER

Fax +41 61 306 12 34
E-Mail karger@karger.ch
www.karger.com

Access to full text and tables of contents, including tentative ones for forthcoming issues: www.karger.com/nps_issues

Editorial

Neuropsychobiology

Neuropsychobiology 2010;62:5–7
DOI: 10.1159/000314304

Published online: May 7, 2010

Lithium: Sixty Years Thereafter

Janusz K. Rybakowski

Department of Adult Psychiatry, Poznan University of Medical Sciences, Poznan, Poland

There is no doubt that in mania patients' improvement has closely paralleled treatment and that this criterion has been fulfilled in the chronic and subacute cases just as closely as in the cases of more recent onset (...) The effect on patients with pure psychotic excitement – that is, true manic attacks – is so specific that it inevitably leads to speculation as to the possible etiological significance of a deficiency in the body of lithium ions in the genesis of this disorder [1].

Sixty years ago, John F. Cade, a psychiatrist from Melbourne, published remarkable results of giving lithium to manic patients in the *Medical Journal of Australia* [1]. This publication may be acknowledged as evidence of the introduction of lithium to modern psychiatric therapy and also as the beginning of modern clinical psychopharmacology as it preceded by 3 years the first French publication concerning the use of an antipsychotic drug, chlorpromazine [2]. It should be noted that the way lithium found its way to psychiatric use would today be interpreted as an overt violation of good clinical practice guidelines and procedures for the introduction of new drugs and that unconventional reasoning and bold actions of John Cade contributed to the fact that today, lithium remains one of the most important psychotropic medications. Fifty-five years after Cade's original publication, a documentary film entitled 'Troubled Minds: The Lithium Revolution' was released, based on interviews with John Cade's sons, his patients and other people connected with him and won the main prize at the International Vega Awards for excellence in scientific broadcasting. Recently, it has even been suggested that the classical form of manic-depressive illness (bipolar I) be named Cade's disease to commemorate the person who introduced the first effective therapeutic remedy for this illness.

The main impact of lithium on bipolar disorder has been exerted by the mood-stabilizing action of this ion, and, actually, lithium has become a prototype of mood-stabilizing drugs. The first evidence for such an effect in the literature has been provided by the English psychiatrist, Geoffrey Hartigan [3], and subsequently by the Danish researchers Paul Baastrup and Mogens Schou [4, 5]. The latter became an indefatigable promoter of lithium use and lithium studies in subsequent years. Mogens Schou is the author of a practical book on lithium treatment for patients and health professionals which had several editions and was translated into many languages [6]. In 1988, Mogens Schou together with Bruno Müller-Oerlinghausen from Germany and Paul Grof from Canada were the founders of the IGSLI – International Group for the Study of Lithium-Treated Patients: recently, the 23rd IGSLI conference took place on October 1–4, 2009 in Halifax, Canada. Since 2001, during biannual conferences of the International Society of Bipolar Disorder held in Pittsburgh, Mogens Schou Awards have been given for exceptional achievements in the field of scientific research, educational activity and organizational and media activity concerning bipolar illness.

The majority of articles in this special issue deal with clinical and neurobiological aspects of the mood-stabilizing effect of lithium. One of them is a paper by Paul Grof [7], a Canadian psychiatrist of Czech origin, entitled

Janusz K. Rybakowski
Department of Adult Psychiatry
Poznan University of Medical Sciences
PL–60572 Poznan (Poland)
Tel. +48 61 847 5087, Fax +48 61 848 0392, E-Mail rybakows@wlkp.top.pl

'Sixty years of lithium responders'. Grof may be remembered for the term 'excellent lithium responders', which he introduced in the mid-1990s. The term denotes the patients on lithium monotherapy in whom the illness virtually ceases to exist [8]. What are clinical characteristics of the excellent lithium responders or, in other words, the people whose bipolar life has been dramatically and successfully changed by lithium? According to Grof, the responders have a characteristic clinical profile with an episodic clinical course, complete remissions, bipolar family history and low psychiatric comorbidity, approximating the classical Kraepelinian description of a manic-depressive illness. The percentage of excellent lithium responders among bipolar patients is about 30%, which has also been confirmed in the study by Rybakowski et al. [9] where patients entering lithium therapy in the 1970s and 1980s were followed for a period of 10 years. However, the most exciting part of Grof's paper is an observation of the second generation of lithium-treated patients. It indicates that parents and offspring suffer from a comparable brain dysfunction that manifests clinically in distinct stages and may have potential for a better understanding and management of mood disorders.

The paper 'The new lithium clinic' was written by Israeli authors under the leadership of Robert Belmaker, describing the evolution of a lithium clinic in the last several decades [10]. According to the experience of the authors, the lithium clinic had originated from a typical outpatient clinic with the aim to assess the serum lithium level and has evolved to a setting incorporating other mood-stabilizing drugs along with tailored psychotherapeutic measures. It should also be mentioned that Robert Belmaker has been a reputable researcher of the lithium mechanism of action and, especially, the connection of the lithium therapeutic effect in bipolar disorder with a phosphatidylinositol (PI) system. Convincing evidence for such an effect was provided in a recent work of his group on Impa1 and Smit1 knockout mice [11]. This may concur with the results of one of the recent genome-wide association studies of bipolar disorder where the strongest association signal among over 550,000 single nucleotide polymorphisms was detected at a marker connected with the PI pathway [12].

Lithium remains a cornerstone in the long-term treatment of bipolar disorder. Despite some doubts expressed in the 1990s as to the quality of placebo-controlled lithium studies performed in 1970–1972 [13], a recent meta-analysis of Geddes et al. [14] fully acknowledges the prophylactic effect of lithium. In this issue, convincing evidence of such a notion based on thorough analysis of the recent literature has also been provided in a paper of Spanish researchers headed by an eminent expert on bipolar disorder, Eduard Vieta [15]. They report data on the comparison of lithium with other drugs with apparently mood-stabilizing action and conclude that lithium should still have a major role in the long-term treatment of bipolar disorder, for preventing both manic and depressive recurrences.

Perhaps the second most important use of lithium nowadays is an augmentation of antidepressants in treatment-resistant depression. This effect is elaborated upon by the Dresden group of researchers under the leadership of Michael Bauer, the current president of the IGSLI [16]. The augmentation of antidepressants by lithium is by far the best evidenced and, apparently, significantly underutilized option for treatment-resistant depression. In the review, it has been pointed out that most studies of lithium augmentation were performed in unipolar depression and included all major classes of antidepressants. The interesting part of the article deals with the recommendation for the duration of lithium administration after successful augmentation and with some pharmacogenomic aspects of such a procedure.

The spectrum of the psychotropic action of lithium goes beyond the mood-stabilizing action. In their mini-review, Bruno Müller-Oerlinhausen and Ute Lewitzka [17] present data pointing out the anti-aggressive and antisuicidal effect of this ion. Evidence for the antisuicidal action of lithium, which so far has not been convincingly demonstrated in mood disorders for any other psychotropic compound, was first obtained in the IGSLI study [18]. The antisuicidal effect of lithium might possibly be related to its anti-aggressive effects which have been shown in various species, populations and settings. Speculations on neurobiological mechanisms of such lithium effects are also provided in the article. Although the antisuicidal effect of lithium is apparently not directly related to the mood-stabilizing action of this ion, it may have a great impact on decreasing the mortality of bipolar patients receiving long-term lithium treatment.

The recent two decades brought about a novel insight into the lithium mechanism of action. The paper written by American investigators, Quiroz et al. [19], headed by a great specialist in the neurobiology of lithium, Husseini Manji, presents the recent data on lithium effects on oxidative stress, apoptosis, inflammation, glial dysfunction, neurotrophic factor functioning, excitotoxicity and mitochondrial stability. The possible targets of lithium action discussed in the paper include cAMP-mediated signal transduction, cAMP response element-binding protein, brain-derived neurotrophic factor, PI cascade, protein kinase C, glycogen synthase kinase 3 and B-cell lymphoma 2. The novel data coming from clinical studies suggesting a neurotrophic effect of lithium are also reviewed. The authors raise hope that a better understanding of the lithium mechanism of action may lead to elaborating improved pharmacological treatment.

In their paper on the pharmacogenetics of lithium, the Italian researchers from Bologna, Alessandro Serretti and Antonio Drago, hypothesize that lithium may drive acute molecular events whose activation over time trigger long-lasting modifications in critical neuronal nets, associated with an expression profile of genes [20]. They discuss the results of pharmacogenetic investigations related to these molecular events. Moreover, they also provide pharmacogenetic data obtained in lithium patients that are not strictly associated with the molecular events thought to be associated with lithium action.

The last paper of this special issue is a product of a truly international initiative of the IGSLI and the National Institute of Mental Health (NIMH). The aim of this enterprise is to perform a genome-wide association study in a substantial international group of lithium-treated bipolar patients. A Consortium on Lithium Genetics (ConLiGen) has been founded under the leadership of Thomas Schulze, a German psychiatrist and geneticist, who is working at the NIMH and is the first author of this article [21]. More than 1,200 patients from several centers all over the world have already been included. A stringent phenotype definition of prophylactic response to lithium is one of the hallmarks of this collaboration. It is hoped that the first results of this great initiative will be available shortly and that they will provide important information for elucidating the lithium mechanism of action.

References

1 Cade JF: Lithium salts in the treatment of psychotic excitement. Med J Aust 1949;36:349–352.
2 Delay J, Deniker P, Harl JM: Utilization en thérapeutique psychiatrique d'une phénothiazine d'action centrale élective. Ann Méd Psychol 1952;110:112–117.
3 Hartigan GP: The use of lithium salts in affective disorders. Br J Psychiatry 1963;109:810–814.
4 Baastrup PC: The use of lithium in manic-depressive psychoses. Compr Psychiatry 1964;5:396–408.
5 Baastrup PC, Schou M: Lithium as a prophylactic agent. Its effect against recurrent depression and manic-depressive psychosis. Arch Gen Psychiatry 1967;16:162–172.
6 Schou M: Lithium Treatment of Mood Disorders. A Practical Guide, ed 6, revised edition. Basel, Karger, 2006.
7 Grof P: Sixty years of lithium responders. Neuropsychobiology 2010;62:8–16.
8 Grof P: Excellent lithium responders: people whose lives have been changed by lithium prophylaxis; in Birch NJ, Gallicchio VS, Becker RW (eds): Lithium: 50 Years of Psychopharmacology. New Perspectives in Biomedical and Clinical Research. Cheshire, Weidner Publishing Group, 1999, pp 36–51.
9 Rybakowski JK, Chlopocka-Wozniak M, Suwalska A: The prophylactic effect of long-term lithium administration in bipolar patients entering treatment in the 1970s and 1980s. Bipolar Disord 2001;3:63–67.
10 Osher Y, Bersudsky Y, Belmaker RH: The new lithium clinic. Neuropsychobiology 2010;62:17–26.
11 Agam G, Bersudsky Y, Berry GT, Moechars D, Lavi-Avnon Y, Belmaker RH: Knockout mice in understanding the mechanism of action of lithium. Biochem Soc Trans 2009;37:1121–1125.
12 Baum AE, Akula N, Cabanero M, Cardona I, Corona W, Klemens B, Schulze TG, Cichon S, Rietschel M, Nöthen MM, Georgi A, Schumacher J, Schwarz M, Abou Jamra R, Höfels S, Propping P, Satagopan J, Detera-Wadleigh SD, Hardy J, McMahon FJ: A genome-wide association study implicates diacylglycerol kinase eta (DGKH) and several other genes in the etiology of bipolar disorder. Mol Psychiatry 2008;13:197–207.
13 Moncrieff J: Lithium revisited. A reexamination of the placebo controlled trials of lithium prophylaxis in manic-depressive disorder. Br J Psychiatry 1995;167:569–574.
14 Geddes JR, Burgess S, Hawton K, Jamison K, Goodwin GM: Long-term lithium therapy for bipolar disorder: systematic review and meta-analysis of randomized controlled trials. Am J Psychiatry. 2004;161:217–222.
15 Nivoli AMA, Murru A, Vieta E: Lithium: still a cornerstone in the long-term treatment in bipolar disorder? Neuropsychobiology 2010;62:27–35.
16 Bauer M, Adli M, Bschor T, Pilhatsch M, Pfennig A, Sasse J, Schmid, Lewitzka U: Lithium's emerging role in the treatment of refractory major depressive episodes: augmentation of antidepressants. Neuropsychobiology 2010;62:36–42.
17 Müller-Oerlinghausen B, Lewitzka U: Lithium reduces pathological aggression and suicidality: a mini-review. Neuropsychobiology 2010;62:43–49.
18 Müller-Oerlinghausen B, Ahrens B, Volk J, Grof P, Grof E, Schou M, Vestergåard P, Lenz G, Simhandl C, Thau K: Reduced mortality of manic-depressive patients in long-term lithium treatment: an international collaborative study by IGSLI. Psychiatry Res 1991;36:329–331.
19 Quiroz JA, Machado-Vieira R, Zarate CA, Manji HK: Novel insights into lithium's mechanism of action: neurotrophic and neuroprotective effects. Neuropsychobiology 2010;62:50–60.
20 Serretti A, Drago A: Pharmacogenetics of lithium long-term treatment. Focus on initiation and adaptation mechanisms. Neuropsychobiology 2010;62:61–71.
21 Schulze TG, Alda M, Adli M, Akula N, Ardau R, Bui ET, Chillotti C, Cichon S, Czerski P, Del Zompo M, Detera-Wadleigh SD, Grof P, Gruber O, Hashimoto R, Hauser J, Hoban R, Iwata N, Kassem L, Kato T, Kittel-Schneider S, Kliwicki S, Kelsoe JR, Kusumi I, Laje G, Leckband SG, Manchia M, MacQueen G, Masui T, Ozaki N, Perlis RH, Pfennig A, Piccardi P, Richardson S, Rouleau G, Reif A, Rybakowski JK, Sasse J, Schumacher J, Severino G, Smoller JW, Squassina A, Turecki G, Young LT, Yoshikawa T, Bauer M, McMahon FJ: The international Consortium on Lithium Genetics (ConLiGen) – an initiative by the NIMH and IGSLI to study the genetic basis of response to lithium treatment. Neuropsychobiology 2010;62:72–78.

Neuropsychobiology

Neuropsychobiology 2010;62:8–16
DOI: 10.1159/000314305

Published online: May 7, 2010

Sixty Years of Lithium Responders

Paul Grof [a, b]

[a] Mood Disorders Center of Ottawa, Ottawa, Ont., and [b] Department of Psychiatry, University of Toronto, Toronto, Ont., Canada

Key Words
Bipolar disorders · Lithium, long-term treatment · Clinical profile · Lithium responders · Offspring · Biological markers

Abstract
It has been 60 years since Cade first described patients who responded to antimanic lithium treatment. Two decades later, responders to lithium stabilization emerged in larger numbers. The responses of many severely ill bipolar patients to lithium were striking and called for an explanation. Remarkable reactions to a simple ion generated hope for an uncomplicated laboratory test of response and an extensive search for suitable biological markers ensued. But despite promising reports, particularly from molecular genetics, we are still waiting for a biological elucidation of the stabilizing effects of lithium. The most useful predictor of lithium stabilization has to date been the patient's clinical profile, based on a comprehensive clinical assessment: complete remissions and other characteristics of episodic clinical course, bipolar family history, low psychiatric comorbidity and a characteristic presenting psychopathology. In brief, the responders approximate the classical Kraepelinian description of a manic-depressive patient. But the most intriguing findings have recently emerged from prospective observations of the next generation: the children of lithium responders, their counterparts coming from parents who did not respond to lithium and controls. Overall, they indicate that parents and offspring suffer from a comparable brain dysfunction that manifests clinically in distinct stages. If the child's predicament starts early in childhood, it presents with varied, nonaffective or subclinical manifestations that are usually nonresponsive to standard treatments prescribed according to the symptoms. The next stage then unfolds in adolescence, first with depressive and later with activated episodes. The observations have a potential to markedly enrich the prevailing understanding and management of mood disorders.

Introduction

The study of response to stabilizing lithium treatment has an interesting history worth revisiting. What we have learned can be applied to clinical practice, incorporated into research on bipolar disorders and may have implications for conceptual thinking about diagnosis and classification of mood disorders. For instance, the clinical profile of bipolar patients with an unequivocal response to long-term lithium treatment (referred to interchangeably as lithium responders or excellent responders) helps to determine who should receive long-term lithium as the treatment of choice. Focusing on patients stabilized on lithium can be effectively used for creating more uniform research samples of bipolar patients. Further, comparing outcomes in unequivocal responders and nonresponders to long-term treatment and in their children is a fertile research strategy for addressing a multitude of questions.

In this article, I will focus on the responders to long-term lithium. A comprehensive review would be too ex-

KARGER

Fax +41 61 306 12 34
E-Mail karger@karger.ch
www.karger.com

0302–282X/10/0621–0008$26.00/0

Accessible online at:
www.karger.com/nps

Paul Grof, MD, PhD, FRCP
Mood Disorders Center of Ottawa, Smyth Medical Center
202-1929 Russell Rd
Ottawa, ON K1G 4G3 (Canada)
E-Mail paul_grof@mdco.ca

tensive, therefore I will concentrate on selected milestones. Lithium stabilization is achieved in a particular type of bipolar patient whose characteristics are outlined below. This effect includes: complete or marked suppression of both manic and depressive recurrences; reliance on therapeutic lithium concentrations; no rebound if lithium treatment is discontinued during full remission; the response reproducible by reinstitution of lithium; toxicity observed only with abnormally elevated lithium levels.

In addition to the well-demonstrated stabilization, lithium may also provide a variety of benefits [1] to other patients: potent antimanic, antipsychotic [2, 3] and antiaggressive [4] effects, antisuicidal and mortality-reducing action [5–7] and antidepressant and antidepressant-augmenting potential. Compared to lithium stabilization, probably the most misunderstood effect of lithium has been its 'antipsychotic' value described best by Garver and his group [2, 3] as well as by others [8, 9]. While this antipsychotic effect of lithium may at times resemble antimanic or stabilizing effects, the responders behave in a qualitatively different way [2, 3, 8, 9], often confusing the clinician. The benefit is not reproducible or lasting and only manic overactivity is affected.

The stabilizing effect of lithium appears to be a unique benefit, and has not been achieved by other pharmacological agents. Unlike in acute treatment, during longitudinal assessment, we can usually determine, with a high probability, if an individual patient is benefitting specifically from the given treatment. The stabilizing effect is therefore methodologically less thorny to research than other lithium-induced changes.

John Cade and the Beginnings

The entry of lithium into clinical psychiatry is associated with the name and work of John Cade, who published the first observations on the successful treatment of acutely ill patients [10]. He was known to be a meticulous observer [11]. Indeed, he provided us with descriptions of lithium responders which, surprisingly, contain seeds of the main features that have been confirmed in subsequent decades.

Cade noted that patients could be effectively treated with lithium not only during acute manias but also successfully maintained subsequently. One of his patients was in fact kept well on lithium successfully over 20 years [12]. Cade also seemed to be aware of some antipsychotic quality of lithium. He actually titled his seminal paper [10] the treatment of 'psychotic states' rather than of mania. Later he also referred to 2 patients suffering from schizophrenia who remarkably benefitted from lithium maintenance [10]. But Cade was also painfully aware of the potential toxicity and side effects associated with lithium use – concerns which seemingly restricted his own use of lithium. In his footsteps, Australian [13] and Danish [13] colleagues later established the principles of monitoring lithium through plasma levels.

To make the story more all-embracing, it should be mentioned that preceding Cade had been earlier reports on the use of lithium in quasi-psychiatric indications [14] but no features of responders were captured. Lange [15] had success with recurrent depressions but, working a century ago, he was deprived of knowing that in fact a large proportion of recurrent depressions manifest a hidden bipolar propensity [16, 17].

The Next Steps

The observations that prompted the testing of the 'prophylactic' use of lithium came from Schou, Baastrup and Hartigan [18–20]. Each of them reported, under different circumstances, that some manic patients who responded to lithium and continued on it remained free of further manias and depressions; a dramatic change compared to the patients' previous histories. Schou and Baastrup [20–22] then pursued these reflections systematically and the rest is history. The stabilizing treatment of bipolar disorder with lithium was born, along with many thousands of responders.

Both the proof of lithium's long-term efficacy and its acceptance in clinical practice were challenging. The idea of pharmacological stabilization was too radical to be accepted easily, particularly as lithium was not a known antidepressant. The methodology of long-term trials still had to be developed [23, 24]. The capricious course of mood disorders had not been properly mapped, so what was thought to happen to untreated patients was based more on speculation than on fact. All this produced fertile soil for the heated debates that ensued [25–27].

But lithium treatment was clearly stabilizing many frequently ill bipolar patients and the efficacy became well documented in a series of double-blind trials [28]. In the 1970s, there was no other competing medication for 'manic-depressive illness' and many patients, often after many years of illness, were eagerly looking for help. To wit, we opened up a lithium clinic in Hamilton, Ont., Canada, in 1968 and in no time had a waiting list of over 1 year long.

Early Successes

A major curiosity about lithium responsiveness emerged after maintenance lithium treatment was finally, after a heated paper war, widely accepted into clinical practice in the early 1970s. However, the pendulum swung and both patients and their physicians often fell under a mistaken impression that lithium was close to a miracle treatment. Many patients with numerous hospitalizations stabilized well for the first time. Their excellent remissions were attributed to the qualities of lithium rather than the nature of episodic illness.

Some bipolar patients responded strikingly well: the term 'excellent lithium responder' was coined [29] for patients who had been markedly ill with frequently recurrent mood disorder and on lithium treatment recovered fully and stayed stabilized. But there were problems. Lithium was introduced as a specific treatment for bipolar disorder [30], yet a fair number of patients not diagnosed as bipolar did benefit from it if tried. Some did temporarily, others in a sustained way; both challenging the assumption of specificity. Conversely, many patients with unquestionable bipolar diagnosis who should have responded failed to do so – the cast of lithium nonresponders. Nagging questions emerged: Why is this happening? What is the difference between responders and treatment failures? How do we identify the responders beforehand?

The real difference between responders and nonresponders could only be clarified by a systematic research. The problem of unexpected, nonbipolar responders was avoided promptly by very liberally widening the diagnostic criteria for mood disorders – a step that hugely complicated all subsequent research of lithium response.

The Impact of Lithium Response on Diagnostic Labeling

Suddenly it seemed that something was not right with the assumptions on which the previous diagnostic systems have been based. It appeared that mood disorders that could perhaps be effectively treated with a new approach had been hidden under other diagnostic tags. It seemed as if, when uncertain, physicians prefer to diagnose that which they feel is treatable. Indeed the clinical statistics demonstrated that, along with acceptance of the effectiveness of lithium, mood disorders suddenly jumped in frequency, while schizophrenia began to disappear [31, 32]. These statistics resembled a mirror image of the vanishing manic-depressive about 15 years earlier [33] when in the minds of some psychiatrists schizophrenia became treatable by new, radical psychotherapeutic approaches.

Interestingly, in years that followed the diagnostic system underwent a basic change. Under the influence of 'neo-Kraepelinians' [34] DSM-III abandoned the earlier theoretical assumptions, and the diagnostic criteria for mood disorders expanded. To diagnose a bipolar disorder it was no more necessary to first exclude abnormal moods as a presentation of other psychiatric disorders; mood-incongruent hallucinations and delusions became an integral part of the bipolar picture, and the attention to the remitting clinical course required by the Kraepelinian approach vanished.

As a result, in this new variety of bipolar patients lithium response rates fell and doubts about lithium's efficacy have kept reemerging periodically, and will continue to do so. Lithium's dramatic success in many patients with classical bipolar illness has justified its unfortunate use in 'bipolar spectrum disorders' and has sown the seeds of disappointment and lithium's relative demise.

Tussle to Find a Biological Marker of Response

Lithium appeared as an effective long-term treatment at the time when biological psychiatry was quickly establishing its dominance over the previous psychoanalytic and psychosocial models. A search for some straightforward biological elucidation took off, quickly and intensely. As many new laboratory techniques were introduced into biological psychiatry, they were also applied to revealing lithium's mode of action and response. But many earlier studies struggled with the methodological complexities of identifying true responders to lithium stabilization and separating them from partial responders and spontaneously remitting false positives.

From a wealth of reports, I have included a few illustrations. Some of the domains of intense search have been lithium transport in red blood cells [35], a red blood cell/plasma lithium ratio [36], urinary lithium excretion [37], platelet monoamine oxidase activity [38], urinary 3-methoxy-4-hydroxyphenylglycol and other expressions of central amine metabolism [39, 40], serum calcium and magnesium [41], average evoked potentials [42, 43] and human leukocyte antigens [44, 45]. As one example of recent studies, in an investigation by Kruger et al. [46] comparing regional cerebral blood flow in valproate responders, lithium responders and their unaffected relatives showed significant, possibly heritable differences between the two patient groups.

We pinned our hopes in the same directions and performed a series of systematic investigations comparing unequivocal responders and nonresponders to lithium stabilization. After correcting for confounding variables, a large set of laboratory findings emerged mostly negative, but a few differences were apparent. We found characteristic neuroendocrine responses in lithium responders [47, 48]. At baseline, healthy controls and symptom-free, remitted responders were indistinguishable, but only responders exhibited significantly different responses after 3 weeks of lithium administered to both groups. These changes in responders are compatible with a serotonin or endorphin trait dysfunction. Lithium responders also differed significantly from healthy and psychiatrically ill controls in the MNS blood groups: 6 versus 24% of NN groups [49].

Thus, many promising biological characteristics of responders and nonresponders were reported but, so far, neither replicated or useful for practical use. The deviations found with new techniques often turned out to reflect incomplete understanding of a multitude of influences on the investigated biological function. Many studies included only animal models or patients treated with lithium only in a routine or brief manner. Very few studies have actually compared carefully classified unequivocal lithium responders with indisputable nonresponders. The optimism of the pilot studies was premature, and the biological marker of lithium response has remained elusive.

Search for Genetic Culprits

The search for genetic markers of lithium response has expanded recently and deserves a separate section. Both bipolar disorder and lithium response cluster in families [50], and the illness has the strongest genetic contribution of all psychiatric ailments. Furthermore, responders to lithium stabilization represent a relatively homogeneous bipolar subtype, quite suitable for genetic research. There have even been recommendations to consider this lithium-responsive classical type of bipolar disorder as an independent illness – perhaps a Morbus Cade [51]. In addition, molecular genetics has already proven its skills and sophistication by recent discoveries in neurology and other branches of medicine.

Several genome scans and meta-analyses have been completed, mostly in international cooperation; others are on the way [52]. The genetic studies of lithium response have offered much promise and I cannot do justice to this exciting, quickly expanding area while staying within the assigned space. I will therefore refer the reader to a recent comprehensive overview [53]. One area not sufficiently covered there is a series of molecular genetic studies of candidate genes in lithium responders carried out in a research program by J. Rybakowski, J. Hauser and coworkers. Their studies include large samples of patients treated with lithium for many years and carefully classified into excellent responders, partial responders and nonresponders. Their positive findings implicate, in particular: serotonin transporter, dopamine D_1 receptor and BDNF genes [54–57].

It seems vital that intense genetic studies of lithium responders continue, as focused efforts on more uniform groups have been successful in other areas of medicine. But it might be too late. It now seems unlikely that a sufficiently sizeable collection of new, unequivocal lithium responders can be gathered for a successful genetic breakthrough. Despite what all bipolar treatment guidelines advise, nowadays very few bipolar patients are systematically stabilized with lithium monotherapy. Drug cocktails dominate the practice of stabilizing treatment, and make it difficult to identify a sufficient number of lithium responders for new genetic studies. If the past is any indication, then the combination of large samples of heterogeneous patients and new laboratory techniques cannot sufficiently compensate for a well-characterized, monotherapy-treated, sufficiently large patient population.

Clinical Profile of Lithium Responders

While we searched extensively for biological explanations of lithium response, the analyses of the accumulated clinical database revealed a clinical profile of excellent responders [58–60].

The features that emerged were later confirmed independently in controlled studies [61, 62] as well as in other clinical and naturalistic studies [63–65]. To summarize the findings: the most important characteristic of the conditions responsive to lithium prophylaxis is the episodic course; that is, episodes of mania and depression alternating with periods of complete remission. During such remissions, the patients not only return to their premorbid functioning in their employment and in their family, they are also free of any residual symptoms, affective or nonaffective. Curiously, the fully remitting, episodic course as a marker of lithium response applies not only to mood disorders (classical or cycloid type of bipolar disorder, episodic recurrent depressions) but also to lithium-responsive medical conditions, such as cluster headaches [66]. A family history of episodic mood disorders and absence of psychiatric comorbidity further increase the probability of successful prophylaxis. Provided the above features from the patient's history are met, the classical symptom presentation in episodes also aids in the prediction of lithium response, such as an intense pervasive sadness or euphoria, changed vital functions such as appetite, sleep and sexual functioning. Mood-congruent psychotic symptoms are also present in some patients.

Patients with this characteristic profile who need stabilizing treatment benefit from lithium as the monotherapy of choice. To assess in practice whether a particular patient with a recurrent mood disorder will stabilize on lithium, it is important to obtain comprehensive information about the patient's clinical profile and carefully interpret it. While the quality of remission is especially important, psychiatrists are usually not trained to assess it and many must first learn to quantify it correctly. An excellent, sustained response to lithium prophylaxis is one of the most gratifying experiences a treating psychiatrist may have.

Studies have shown that lithium is and remains highly effective for the treatment of classical bipolar disorder which dominated the original clinical trials [67, 68]. In clinical trials with broadly diagnosed, heterogeneous bipolar spectrum disorders the efficacy of lithium has varied from low to impossible to demonstrate [69].

Three recent studies focusing on detecting the clinical predictors of lithium response require a more detailed treatise [69–71]. Kleindienst et al. [70] selected 43 suitable clinical trials of lithium from all relevant papers published between 1966 and 2003. Data were analyzed by meta-analytic aggregation based on a random effects model and effect size measure. Of 42 clinical variables, an episodic pattern of MDI (mania-depression-interval sequence), and a higher age at illness onset were associated with response, while a high number of previous hospitalizations, an episodic pattern of DMI (depression-mania-interval sequence) and a pattern of continuous cycling worsened the chances of response. For another 6 variables significantly related to response the evidence was not deemed sufficiently robust. The authors concluded that none of the investigated variables actually had a strong impact on lithium response which seemed to be related to a multitude of variables.

The strength of this study is in the size of gathered material and sophisticated statistical analysis. Unfortunately, the poor suitability of the analyzed data, including the heterogeneity of the included studies, is illustrated well in the contradictory findings. When investigating the same variables, such as frequency of episodes, age at onset and bipolar subtype, in different studies, the relationship to response was sometimes positive, and at other times negative or nonexistent. The effect sizes calculated for other variables, such as duration of illness, different types and length of index episodes, are characterized by the authors as having 'substantial', 'enormous' or 'extreme' variation across studies. Other crucial questions about this report are the reliability of the predictors identified and in particular interdependence of the predictors. Unfortunately, observations from drug trials may suggest some association with response, but they are unable to provide proof of response in individuals. This review can only be considered as a preliminary step toward a comprehensive analysis.

However, our findings support the authors' when they state that past history of a patient seems to predict response to prophylactic lithium better than the patient's presenting clinical picture and that, at the current stage of knowledge, it seems preferable to base the prognosis on a multitude of variables, rather than on any single one.

Two analyses of the material gathered by the IGSLI (International Group for the Study of Lithium-Treated Patients) also deal with the clinical characteristics of lithium response. The data summarize the clinical profile and outcome of 336 bipolar I and II patients from 5 IGSLI centers who initially responded to lithium treatment and then continued on it for up to 30 years. The analysis by Pfennig et al. [71] addresses to what extent the response to long-term lithium was influenced by clinical features. Conditional and frailty regression models, as well as a combined model were applied and the outcome assessed by the time elapsed to recurrence. The risk of recurrence increased significantly in the presence of residual symptoms in between episodes, mood-incongruent psychotic symptoms during the episodes and a rapid cycling course.

I was given the opportunity to statistically reanalyze the same data set. To stay close to clinical practice, as the overall outcome indicator I used the Alda score [50] from the database. High scores on the Alda scale indicate a high probability that the patient is doing well because of the administered lithium. To optimize the predictor detection, I employed the discriminant function analysis. In agreement with Pfennig's findings, the clinical features such as the quality of remission, absence of mood-incongruent psychotic symptoms and family history of bipolar disorder were highly significant predictors (p <

0.001). The high significance level was presumably related to the large sample size. But the variables together explained only a small proportion of variance as this was, by definition, a sample of patients predominantly responding well to lithium. The high interdependence of the variables was striking and the quality of remission accounted for the majority of explainable variance.

In the analysis by Berghoefer et al. [72], data from a subsample of 242 IGSLI patients were employed. Typical and atypical features were documented on an 8-item scale. It is important that each feature was handled as having the same importance, as independent with the same weight. In 142 patients, the number of typical features was greater than the number of atypical features, whereas in 100 patients the number of atypical features was greater than or equal to the number of typical features. Statistical analyses showed that the stability of response to long-term lithium treatment persists over time and there is no evidence of a loss of efficacy as the treatment progressed, regardless of clinical features. The predominance of either typical or atypical features did not result in different responses to long-term lithium treatment in this sample of bipolar patients.

Based on statistical reanalyses, my interpretation of these two studies is as follows. These longitudinal data were well suited for testing the stability of long-term effect versus its possible loss, and demonstrated that in bipolar patients with mainly classical profile the response is stable over time, regardless of the patient's clinical features. Because the collection favored better-responding patients over nonresponders, the material was less suited for identifying the forecasters of response. Despite that, when tested with models that account for the interdependence of clinical features, the predictors that emerged are compatible with previous clinical predictors. On the other hand, if one strived for a bimodal predictor (such as predominance of typical or atypical features), one would have to consider the interdependence of included variables.

Second Generation of Responders

Prior to the availability of stabilization on lithium, the usual advice to women with bipolar disorder was not to have children, in order not to propagate the predicament onto the next generation. When studying the effect of pregnancy on the course of bipolar illness [73], we were surprised at the large proportion of affected women who actually had complied and avoided becoming pregnant. The success of lithium seems to have strikingly changed this. As women who for the first time in their life experienced stabilization on lithium became parents, they started bringing their troubled children to us for assessment and care. As we had helped the adults, parents were assuming that we would be able to do the same for their children.

This was a new field, and treating these children's attacks of anxiety, phobias and obsessions the standard way, according to their symptoms, rarely worked. We were initially hesitant to use lithium, for several reasons, but on a few occasions when we dared, lithium helped. We made some other interesting pilot observations: for example, we knew from long-term observations of affected adults that acute episodes of illness are accompanied by a positive dexamethasone suppression test, reflecting heightened brain arousal. Similarly in their children, the dexamethasone suppression test also turned positive during the early, nonspecific manifestations of illness – during the episodes of acute anxiety, panic, phobias or obsessions – and normalized afterwards.

Then, in 1994, a young courageous psychiatrist joined our program and decided to find out what was really going on with the children of bipolar parents. A series of prospective 'high-risk' studies of offspring of bipolar parents was born, spearheaded by Dr. Anne Duffy [74–78]. The enrolled parents all participated in the mood disorders research program and thus their clinical profile was well captured. To qualify, they had to be either an unequivocal responder or nonresponder to stabilizing lithium treatment.

The high-risk studies were designed such that the children were assessed by an independent team blinded to the parent's clinical profile, and then followed prospectively whenever they become symptomatic, and otherwise annually. Full information about the child's infant years was recovered from each parent. A control group of children born of parents free from psychiatric illness was also included. Such a longitudinal investigation of offspring of bipolar parents is a fruitful strategy for charting the beginnings of the illness, identifying the sources of vulnerability and protection, uncovering the interplay of important pathophysiological factors and understanding treatment response.

At this point, 212 high-risk children have been followed, and more are being added. The findings have markedly enriched those of the parents. Over the course of these longitudinal investigations, Duffy has made a novel discovery that bipolar disorder evolves in stages. Childhood sleep and anxiety disorders develop around puberty into minor mood disturbances and sensitivity to stress. This is followed by the onset of major depressive episodes, starting in mid-adolescence, while activated mood episodes – hypomanic or manic – follow several years later. The illness may begin at any of these stages in individuals, but then usually follow the identified sequence. Offspring of responders develop episodic disorders, while offspring of nonresponders show either a partially remitting or nonremitting fluctuating course of illness.

Duffy has also shown that early in the course of illness offspring who develop bipolar disorder can be stabilized on monotherapy and that associated predictors of response include the nature of the clinical course (episodic vs. nonepisodic) and the response in the parent. These observations support the classical literature in regard to predictors of lithium response already described and suggest the importance of including careful clinical assessment of the clinical course and family history of disorders and treatment response into the diagnostic and treatment formulation of individual patients. In other studies, we have shown that neurocognitive, neurochemical and neuroanatomical findings in patients with established illness do not appear as vulnerability traits in the offspring, but are related to the burden of illness – emphasizing the importance of early detection and effective intervention.

This landmark high-risk study appears to be the largest and longest of its type. It has already provided important clinical information relevant to early identification and treatment of bipolar disorder and helped to reconcile controversies. The research now proceeds to integrate findings from clinical, genetic, biological and psychosocial research in one bipolar model.

These intriguing findings also alter significantly the previous thinking about the onset of these disorders. When interviewed about their childhood, adult lithium responders describe their prepubertal childhood as completely free of any psychiatric problems [79]. But many of the high-risk children of lithium responders studied prospectively show psychiatric problems already in childhood. Yet, though the psychopathology is qualitatively different, there are good reasons to believe that these psychopathological manifestations are actually arising from the same brain dysregulation. First, the characteristics of the clinical course mirror the ill parent; second, the stages progress later into manifestations of a mature bipolar disorder, and third, the children appear to respond to the same type of mood stabilizer as their parent, rather than what one would expect according to their symptoms.

Similar to the reports about high-risk prepsychotic children [80], these findings challenge some of the long-standing assumptions of diagnosing and classifying primarily on the basis of psychopathological symptoms. More important, they might help explaining many of the prevailing treatment frustrations and failures.

Concluding Remarks

It has been decades since Cade first described responders to lithium during antimanic treatment and Schou and Baastrup through stabilization. The subsequent search for biological markers of response to a simple lithium ion has turned out surprisingly complex. Even after several decades of intense explorations, the expedition still does not seem to have moved far from base camp. Recent genetic studies offer more promise nevertheless have no genes in hand, as yet. One of the main obstacles to research has been correct identification of patients as responders and nonresponders to long-term lithium. This is particularly challenging for bipolar disorder as natural course is capricious, individually variable and patients waver in their compliance with long-term medication. Overreliance on laboratory has also complicated the matter; the most highly technologically evolved measure applied to a mixed clinical population will not succeed.

The most useful indicator of future success with lithium stabilization is the clinical profile of patient. The fully remitting course of illness is the strongest predictor, particularly when accompanied by family history of episodic bipolar disorders and no psychiatric comorbidity. The clinical picture of lithium-responsive bipolar subtype resembles the classical Kraepelinian description of bipolar disorder.

The most intriguing findings that have recently emerged come from the observations of the next generation: from prospective studies of the children of bipolar responders, their counterparts of nonresponsive parents and controls. These studies have uncovered that bipolar disorders evolve in stages that often start with nonspecific manifestations in childhood. They also provide the opportunity to research the onset and progression of illness uncontaminated yet by burden of illness and multiple treatments.

There has recently been partial renaissance of interest in lithium treatment, and in its possible modes of action, picking up where we left off. The study of lithium responders, nonresponders and their offspring plays an important role in this line of research which promises to uncover a new, more effective and safer way of stabilizing bipolar patients.

Acknowledgements

The studies could have never been completed without major contributions from my colleagues Anne Duffy (Ottawa/Halifax) and Martin Alda (Halifax) and from senior members of the IGSLI group, in alphabetical order, Anne Berghoefer (Berlin), Eva Grof (Ottawa), Jan and Eva Libiger (Hradec), Rasmus Licht (Aarhus), Bruno Muller-Oerlinghausen (Berlin), Agneta Nielsen (Stockholm), Janusz Rybakowski (Poznan), Mogens Schou (Aarhus), Ola Suwalska (Poznan), Ken Thau (Vienna), Per Vestergaard (Aarhus), Milos Vojtechovsky and Petr Zvolsky (both Prague). It was extremely valuable that the members of the IGSLI allowed me to interview, test and review their patients and shared their data.

References

1 Grof P: Has the effectiveness of lithium changed? Impact of the variety of lithium's effects. Neuropsychopharmacology 1998;19: 183–188.
2 Garver DL, Hirschowitz J, Fleishmann R, Djuric PE: Lithium response and psychoses: a double-blind, placebo-controlled study. Psychiatry Res 1984;12:57–68.
3 Garver DL, Hutchinson LJ: Psychosis, lithium-induced antipsychotic response, and seasonality. Psychiatry Res 1988;26:279–286.
4 Sheard MH: The effect of lithium and other ions on aggressive behavior. Mod Probl Pharmacopsychiatry 1978;13:53–68.
5 Muller-Oerlinghausen B, Muser-Causemann B, Volk J: Suicides and parasuicides in a high-risk patient group on and off lithium long-term medication. J Affect Disord 1992; 25:261–270.
6 Muller-Oerlinghausen B, Felber W, Berghoefer A, Lauterbach E, Ahrens B: The impact of lithium long-term medication on suicidal behavior and mortality of bipolar patients. Arch Suicide Res 2005;9:307–319.
7 Tondo L, Jamison KR, Baldessarini RJ: Effect of lithium maintenance on suicidal behavior in major mood disorders. Ann NY Acad Sci 1997;836:339–351.
8 Kingsbury SJ, Garver DL: Lithium and psychosis revisited. Prog Neuropsychopharmacol Biol Psychiatry 1988;22:249–263.
9 Lenz G, Lovrek A, Thau K, Topitz A, Denk A, Simhandl C, Wancata J, Wolf R: Lithium prophylaxis of schizoaffective disorders (in German). Wien Med Wochenschr 1987;137:5.
10 Cade J: Lithium salts in the treatment of psychotic excitement. Med J Aust 1949;36:349–352.
11 Johnson FN: The History of Lithium Therapy. London, The MacMillan Press Ltd, 1984.
12 Cade JFJ: Lithium – Past, present and future; in Johnson FN, Johnson S (eds): Lithium in Medical Practice. Baltimore, University Park Press, 1978.
13 Coats DA, Trautner EM, Gershon S: Treatment of lithium poisoning. Aust Ann Med 1957;1:11–15.
14 Haig A: Mental depression and the excretion of uric acid. Practitioner 1988;41:342–354.
15 Lange C: Periodische Depressionszustände und ihre Pathogenesis. Hamburg, Leopold Voss, 1896.
16 Akiskal H, Bourgeois M, Angst J, Post R, et al: Re-evaluating the prevalence of and diagnostic composition within the broad clinical spectrum of bipolar disorders. J Affect Disord 2000;59:S5–S30.
17 Angst J, Gamma A: Prevalence of bipolar disorders: traditional and novel approaches. Clin Appr Bipol Disord 2002;1:10–14.
18 Baastrup PC: The use of lithium in manic-depressive psychosis. Compr Psychiatry 1963;5:396–408.
19 Hartigan GP: The use of lithium salts in affective disorders. Br J Psychiatry 1963;109: 810–814.
20 Baastrup PC, Schou M: Lithium as a prophylactic agent: its effect against recurrent depressions and manic-depressive psychosis. Arch Gen Psychiatry 1967;16:162–172.
21 Baastrup PC, Poulsen JC, Schou M: Prophylactic lithium: double blind discontinuation in manic-depressive and recurrent-depressive disorders. Lancet 1970;ii:326.
22 Schou M, Baastrup PC, Grof P, et al: Pharmacological and clinical problems of lithium prophylaxis. Br J Psychiatry 1970;116:615–619.
23 Grof P, Schou M, Angst J, Baastrup PC, Weis P: Methodological problems of prophylactic trials in recurrent affective disorders. Br J Psychiatry 1970;116:599–603.
24 Grof P: Designing long-term clinical trials in affective disorders. J Affect Disord 1994;30: 243–255.
25 Blackwell B, Shepherd M: Prophylactic lithium: another therapeutic myth? An examination of the evidence to date. Lancet 1968;i:968–971.
26 Blackwell B: Lithium: prophylactic or panacea? Med Counterpoint 1969;11:52–59.
27 Angst J, Grof P, Schou M: Lithium. Lancet 1969;i:1097.
28 Schou M, Thomsen K: Lithium prophylaxis of recurrent endogenous affective disorders; in Johnson FR (ed): Lithium Research and Therapy. New York, Academic Press, 1975.
29 Grof P: Excellent lithium responders: people whose lives have been changed by lithium prophylaxis; in Birch NJ, Gallicchio VS, Becker RW (eds): Lithium: 50 Years of Psychopharmacology. Cheshire, Weidner Publishing Group, 1999.
30 Schou M, Thomsen K, Baastrup PC: Studies on the course of recurrent endogenous affective disorders. Int Pharmacopsychiatry 1970; 5:100–106.
31 Baldessarini RJ: Frequency of diagnosis of schizophrenia versus affective disorders from 1944 to 1968. Am J Psychiatry 1970; 127:759–763.
32 Grof P, Fox D: Admission rates and lithium therapy (letter to the editor). Br J Psychiatry 1987;150:264–265.
33 Lehmann HE: Clinical features of schizophrenia; in Freedman AM, Kaplan H (eds): Comprehensive Textbook of Psychiatry. Baltimore, Williams and Wilkins, 1967.
34 Klerman G: Historical perspectives on contemporary schools of psychopathology; in Millon T, Klerman GL (eds): Contemporary Directions in Psychopathology: Toward the DSM-IV. New York, Guilford Press, 1986.
35 Upadhyaya AK, Varma VK, Sankaranayanaran A: Lithium in prophylactic therapy of manic-depressive illness: biochemical correlates of response. Biol Psychiatry 1985; 20:202–205.
36 Mendels J, Fraser A, Baron J: Intra-erythrocyte lithium ion concentration and long-term maintenance treatment. Lancet 1976;i:966.
37 Serry M: The lithium excretion test. Aust NZ J Psychiatry 1969;3:390–394.
38 Sullivan JL, Cavenar JO, Maltbie A: Platelet-monoaminooxidase activity predicts response to lithium in manic-depressive illness. Lancet 1977;ii:1325–1327.
39 Carroll BJ: Prediction of treatment outcome with lithium; in Cooper TN, Gershon S, Kline N, et al (eds): Lithium, Controversies and Unresolved Issues. Amsterdam, Excerpta Medica, World Congress Lecture Series, 1979.
40 Goodwin FK, Post R, Dunner DL, Gordon EK: Cerebrospinal amine metabolites in affective illness. Am J Psychiatry 1973;130:73–79.
41 Carman JS, Post RH, Teplitz TA, Goodwin FK: Divalent cations in predicting antidepressant response to lithium. Lancet 1974;ii:1454.
42 Baron M, Gershon ES, Rudy V, Jonas WZ, Buchsbaum M: Lithium carbonate response in depression. Arch Gen Psychiatry 1975;32: 1107–1111.
43 Buchsbaum M, Goodwin F, Murphy D: AER in affective disorders. Am J Psychiatry 1971; 128:19–25.

44 Maj M: The impact of lithium prophylaxis on the course of bipolar disorder: a review of the research evidence. Bipolar Disord 2000;2: 93–101.
45 Maj M, Arena F, Loverno N, et al: Factors associated with response to lithium prophylaxis in DSM III major depression and bipolar disorder. Pharmacopsychiatry 1985;18: 309–313.
46 Kruger S, Alda M, Young LT, Goldapple K, Parikh S, Mayberg H: Risk and resilience markers in bipolar disorder: brain responses to emotional challenge in bipolar patients and their healthy siblings. Am J Psychiatry 2006;163:257–264.
47 Grof E, Grof P, Brown GM, Downie S: Lithium effects on neuroendocrine function. Prog Neuropsychopharmacol Biol Psychiatry 1984;8:541–546.
48 Grof E, Grof P, Brown GM: Investigations of lithium effects on neuroendocrine function in man; in Birch NJ (ed): Lithium: Inorganic Pharmacology and Psychiatric Use. Washington, IRL Press, 1987.
49 Alda M, Grof P, Grof E: MN blood groups and bipolar disorder: evidence of genotypic association and Hardy-Weinberg disequilibrium. Biol Psychiatry 1998;44:361–363.
50 Grof P, Duffy A, Cavazzoni P, Grof E, Garnham J, MacDougall M, O'Donovan C, Alda M: Is response to prophylactic lithium a familial trait? J Clin Psychiatry 2002;63: 942–947.
51 Ghaemi SN, Ko J, Goodwin FK: 'Cade's disease' and beyond: misdiagnosis, antidepressant use, and a proposed definition for bipolar spectrum disorder. Can J Psychiatry 2002;47:125–134.
52 Schulze TG, Alda M, Adli M, Akula N, Ardau R, Bui ET, Chillotti C, Cichon S, Czerski P, Del Zompo M, Detera-Wadleigh SD, Grof P, Gruber O, Hashimoto R, Hauser J, Hoban R, Iwata N, Kassem L, Kato T, Kittel-Schneider S, Kliwicki S, Kelsoe JR, Kusumi I, Laje G, Leckband SG, Manchia M, MacQueen G, Masui T, Ozaki N, Perlis RH, Pfennig A, Piccardi P, Richardson S, Rouleau G, Reif A, Rybakowski JK, Sasse J, Schumacher J, Severino G, Smoller JW, Squassina A, Turecki G, Young LT, Yoshikawa T, Bauer M, McMahon FJ: The international Consortium on Lithium Genetics (ConLiGen) – an initiative by the NIMH and IGSLI to study the genetic basis of response to lithium treatment. Neuropsychobiology 2010;62:72–78
53 Alda M, Grof P, Zvolsky P: Genetic factors and response to lithium treatment; in Bauer M, Grof P, Mueller-Oerlinghausen B (eds): Lithium in Neuropsychiatry: The Comprehensive Guide. London, Informa, 2006.
54 Rybakowski JK, Suwalska A, Skibinska M, Szczepankiewicz A, Leszczynska-Rodziewicz A, Permoda A, Czrski PM, Hauser J: Prophylactic lithium response and polymorphism of the brain-derived neurotrophic factor gene. Pharmacopsychiatry 2005;38: 186–170.
55 Rybakowski JK, Suwalska A, Czrski PM, Dmitrzak-Weglarz M, Leszczynska-Rodziewicz A, Hauser J: Prophylactic effect of lithium in bipolar affective illness may be related to serotonin transporter genotype. Pharmacol Rep 2005;57:124–127.
56 Rybakowski JK, Suwalska A, Skibinska M, Dmitrzak-Weglarz M, Leszczynska-Rodziewicz A, Hauser J: Response to lithium prophylaxis: interaction between serotonin transporter and BDNF genes. Am J Med Genet B Neuropsychiatr Genet 2007; 144B:820–823.
57 Rybakowski JK, Dmitrzak-Weglarz M, Suwalska A, Leszczynska-Rodziewicz A, Hauser J: Dopamine D_1 receptor gene polymorphism is associated with prophylactic lithium response in bipolar disorder. Pharmacopsychiatry 2009;42:20–22.
58 Grof P, Lane J, MacCrimmon D, Werstiuk E, Blajchman M, Daigle L, Varma R: Clinical and laboratory correlates of the response to long-term lithium treatment; in Stromgren E, Schoua M (eds): Origin, Prevention and Treatment of Affective Disorders. London, Academic Press, 1979.
59 Grof P, Hux M, Grof E, Arato M: Prediction of response to stabilizing lithium treatment. Pharmacopsychiatry 1983;16:195–200.
60 Grof P, Alda M, Grof E, Fox D, Cameron P: The challenge of predicting response to stabilising lithium treatment. The importance of patient selection. Br J Psychiatry Suppl 1993;163(suppl 21):16–19.
61 Passmore M, Garnham J, Duffy A, Munro A, Slaney C, Teehan A, Alda M: Phenotypic spectra of bipolar disorder in responders to lithium versus lamotrigine. Bipolar Disord 2003;5:110–114.
62 Greil W, Kleindienst N, Erazo N, Müller-Oerlinghausen B: Differential response to lithium and carbamazepine in the prophylaxis of bipolar disorder. J Clin Psychopharmacol 1998;18:455–460.
63 Abou-Saleh MT, Coppen A: Who responds to prophylactic lithium? J Affect Disord 1986;10:115–125.
64 Abou-Saleh MT: Predictors of response to prophylactic lithium; in Birch NJ (ed): Lithium: Inorganic Pharmacology and Psychiatric Use. Washington, IRL Press, 1987.
65 Maj M, Del Vecchio M, Starace F, Pirozzi R, Kemali D: Prediction of affective psychoses response to lithium prophylaxis. Acta Psychiatr Scand 1984;69:37–44.
66 Damasio H, Lyon L: Lithium carbonate in the treatment of cluster headaches. J Neurol 1980;224:1–8.
67 Berghoefer A, Kossman AB, Mueller-Oerlinghausen B: Course of illness and pattern of recurrences in patients with affective disorders during long-term lithium prophylaxis: a retrospective analysis over 15 years. Acta Psychiatr Scand 1996;93:349–354.
68 Rybakowski J, Chlopocka-Wozniak M, Suwalska A: The prophylactic effect of long-term lithium administration in bipolar patients entering treatment in the 1970s and 1980s. Bipolar Disord 2001;3:63–67.
69 Bowden CL, Calabrese JR, McElroy SL, Gyulai L, Wassef A, Petty F, Pope HG Jr, Chou JCY, Keck PE, Rhodes LJ, Swann AC, Hirschfeld RMA, Wozniak PJ, for the Divalproex Maintenance Study Group: A randomized, placebo-controlled 12-month trial of divalproex and lithium in treatment of outpatients with bipolar I disorder. Arch Gen Psychiatry 2000;57:481–489.
70 Kleindienst N, Engel R, Greil W: Which clinical factors predict response to prophylactic lithium? A systematic review for bipolar disorders. Bipolar Disord 2005;7:404–417.
71 Pfennig A, Schlattmann P, Alda M, Grof P, Glenn T, Muller-Oerlinghausen B, Suwalska A, Rybakowksi JK, Willich S, Bauer M, Berghoefer A: Clinical variables influencing the quality of prophylactic effectiveness of long-term lithium treatment in bipolar patients. Submitted.
72 Berghoefer A, Alda M, Mazda A, Baethge Ch, Bauer M, Bschor T, Glenn T, Grof P, Muller-Oerlinghausen B, Rybakowksi JK, Suwalska A, Pfennig A: Long-term effectiveness of lithium in bipolar disorder: a multicenter investigation of patients with typical and atypical features. J Clin Psychiatry 2008; 69:1860–1868.
73 Grof P, Robbins W, Alda M, Berghoefer A, Vojtechovsky M, Nilsson A, Robertson C: Protective effect of pregnancy in women with lithium-responsive bipolar disorder. J Affect Disord 2000;61:31–39.
74 Duffy A, Alda M, Kutcher S, Fusee C, Grof P: Psychiatric symptoms and syndromes among adolescent children of parents with lithium-responsive or lithium-nonresponsive bipolar disorder. Am J Psychiatry 1998; 155:431–433.
75 Duffy A: A prospective study of the offspring of bipolar parents responsive and nonresponsive to lithium treatment. J Clin Psychiatry 2002;63:1171–1178.
76 Duffy A: Does bipolar disorder exist in children? A selected review. Can J Psychiatry 2007;52:409–417.
77 Duffy A, Alda M, Milin R, Grof P: A consecutive series of treated affected offspring of parents with bipolar disorder: is response associated with the clinical profile? Can J Psychiatry 2007;52:369–375.
78 Duffy A, Alda M, Hajek T, Grof P: Early course of bipolar disorder in high-risk offspring: prospective study. Br J Psychiatry 2009;195:457–458.
79 Grof P: Selecting effective treatments for bipolar disorders. J Clin Psychiatry 2003; 64(suppl 5):53–61.
80 McGorry PD, Killackey E, Yung A: Early intervention in psychosis: concepts, evidence and future directions. World Psychiatry 2008;7:148–156.

Neuropsychobiology

Neuropsychobiology 2010;62:17–26
DOI: 10.1159/000314306

Published online: May 7, 2010

The New Lithium Clinic

Yamima Osher Yuly Bersudsky R.H. Belmaker

Beer Sheva Mental Health Center, Ministry of Health and Ben Gurion University of the Negev, Beer Sheva, Israel

Key Words
Bipolar disorder · Lithium clinic · Treatment model · Psychotherapy

Abstract
Until the early 1950s, no effective pharmacological treatment existed for bipolar affective disorder. By the early 1960s, specialty clinics were being set up to dispense lithium carbonate to bipolar patients. By the late 1980s, a new body of knowledge was influencing the perception of bipolar disorder and how the disease should be treated. The authors' lithium clinic from 1974 has grown and evolved from a lithium blood level monitoring model into a comprehensive care model with polypharmacy, psychoeducation, rehabilitation, cognitive therapy, social rhythm therapy, and employment counseling as well as a staff of 2 part-time psychiatrists and 1 clinical psychologist. This service delivery model may benefit both treatment and research in bipolar disorder. The evolution of psychopharmacological and psychosocial knowledge in treating bipolar illness has been integrated into our clinic. Case vignettes are presented to illustrate these points. The comparative cost of this model is discussed.

Introduction: Historical Background

Report of the prophylactic efficacy of lithium caused excitement and a flurry of research into this simple but powerful ion [1]. By the early 1960s, specialty clinics were being set up explicitly to dispense the drug. One of the first of these 'lithium clinics' was established in New York by Fieve [2]. Kerry in Sheffield, UK [3] and a group of private practitioners in Normal, Ill., USA [4] published early reports of similar 'lithium clinics'. In the late 1970s, the name 'lithium clinic' continued to be popular [5]. These early clinics functioned in a way designed to maximize efficiency, especially regarding physician's hours, based on the perception that the blood level of lithium was the main target of intervention. Patients were often seen by a nurse, individually or in a group, or by paramedical personnel who filled out rating scales. Blood levels were taken, and typically the physician was called in 'only if the patient requires a change in dosage, is experiencing mood change, or shows possible symptoms of toxicity' [2]. In one clinic, if the patient was unable to attend, blood samples were simply sent in by the local health authority [3]. As late as the 1990s, clinicians who had worked with lithium clinics could write how their experiences had 'impressed upon us the potency and effectiveness of lithium, and its ability to transform patients' lives from disastrous to manageable and normal' [6].

KARGER

Fax +41 61 306 12 34
E-Mail karger@karger.ch
www.karger.com

0302–282X/10/0621–0017$26.00/0

Accessible online at:
www.karger.com/nps

Yamima Osher
Ministry of Health, Beer Sheva Mental Health Center, Faculty of Health Sciences
Ben Gurion University of the Negev
PO Box 4600, Beer Sheva 84170 (Israel)
Tel. +972 8 640 1519, Fax +972 8 640 1621, E-Mail yamy@bgu.ac.il

A different model developed in Berlin in the early 1980s, where the lithium clinic annually treated about 100 mood disorder patients [7]. In this clinic, which saw each patient an average of 8 times per year, the patient was seen by a psychiatrist at each visit, not only by a nurse or social worker. This multidisciplinary approach emphasized talking to the patients about their problems, not only checking compliance and filling out new prescriptions. The social worker assisted the patient in finding employment and living arrangements, and intervened in crises [7].

By the late 1980s, a growing body of knowledge was changing the perception of bipolar disorder overall. The authors' lithium clinic began in 1974 in Jerusalem, moved to Beer Sheva in 1985, employed a psychologist in 1990 and a second psychiatrist in 2000. From 1974 to 1985, one of the authors (R.H.B.) had an EEL flame photometer [8] in his office and performed serum lithium determinations himself while patients waited. At present, the clinic is a regional specialty clinic serving approximately 180 bipolar I patients. While the catchment area of our clinic includes 500,000 residents, patients also have access to private care or to a psychiatry clinic in the general medical hospital. They come to our University Bipolar Clinic voluntarily. Usually referral is at discharge from one of the inpatient wards (the only inpatient service in our catchment area) and for this reason our clinic is limited to bipolar I patients. Patients are aware of the advantage of continuity of care that our clinic offers and that, on the other hand, the likelihood exists of being approached for participation in research trials. Some complain of long waiting times or the lack of waiting room privacy. Solutions to this last issue have not been found consistent with our large patient volume and cost considerations (see below). This paper will describe the evolution of both psychopharmacological and psychosocial approaches to treating bipolar illness in our clinic, and how these changes have been integrated into clinical practice. Case vignettes from our clinic are presented to illustrate particular points and the cost of this model will be discussed.

Pharmacotherapy

New developments in the psychopharmacology of bipolar disorder have been reviewed extensively in recent years [9, 10]. It is increasingly recognized that the average bipolar patient takes several medications with the goal of reducing to a minimum the number of manic and depressive recurrences, even though monotherapy trials versus placebo show effectiveness for each of the compounds [10, 11]. The first compound to be reported as effective in addition to lithium in the prophylaxis of bipolar disorder was carbamazepine by Okuma et al. [12] in Japan, although the work was greatly extended and given scientific impact in both basic and clinical studies by Post et al. [13]. Carbamazepine was added to the therapeutic armamentarium in our clinic in the early 1980s [14]. At a meeting in Basel in 1989 relating to the possible development of oxcarbazepine, internal documents of Ciba Geigy, which was then the exclusive supplier of carbamazepine, showed that worldwide off-label sales of carbamazepine for bipolar disorder represented about half of worldwide sales of the anticonvulsant. Few controlled studies were done, however, and only recently was a long-acting carbamazepine preparation FDA approved for bipolar disorder.

Valproic acid had similarly been reported to be effective in bipolar disorder in Europe by Emrich et al. [15], but the major impact and acceptance of this medicine came after controlled American trials by Bowden et al. [16]. Valproate began to be widely used in our clinic by the late 1990s. The divalproex sodium preparation was never used, only generic valproic acid. More recently, lamotrigine was FDA approved for bipolar disorder and its use has been reviewed [17].

While the fact that several anticonvulsants are effective in bipolar disorder would suggest a relationship between seizure disorder and bipolar disorder, no pathophysiological evidence has been proven to support this hypothesis. It would appear that anticonvulsants that act on the sodium channel are usually effective in bipolar disorder, whereas anticonvulsants such as phenobarbital or clonazepam that act on the benzodiazepine receptor are ineffective in bipolar disorder [18, 19]. Sometimes phenytoin was used as a counterexample of a good anticonvulsant that is not effective in bipolar disorders; however, our own studies suggest that phenytoin may indeed be effective [20]. Topiramate was studied for bipolar disorder but failed in large clinical trials. This failure may have been due to the large number of antidepressant-induced manias in those studies that led to a corresponding high response rate of the placebo group. The fact that topiramate is not approved as an antibipolar drug does not prove the opposite, i.e., that it is not an antibipolar drug. This distinction is very important in deciding what to do for a patient who has not responded to any other treatment and topiramate has been occasionally useful.

Our concept of the specificity of pharmacological treatment of bipolar disorder is illustrated in figure 1. It would be scientifically elegant if we could divide bipolar patients into (1) those that are lithium responsive and unresponsive to anticonvulsants and (2) those that are lithium unresponsive and responsive to anticonvulsants. If this were true, it would be useful in further genetic studies of bipolar disorder as well as in pathophysiology. However, this is not the case clinically: we had patients who were responsive to both lithium and anticonvulsants as well as patients who were unresponsive to both lithium and to anticonvulsants. These clinical facts become evident when a patient must stop lithium because of side effects and go over to an anticonvulsant and vice versa. There clearly are some patients who are uniquely responsive to lithium but many quite typical bipolar patients who had to stop lithium have done equally well on valproate or carbamazepine.

Many atypical antipsychotics have been shown to be effective in several phases of bipolar disorder and are recommended in guidelines to this effect [21], particularly quetiapine, ziprasidone, olanzapine, risperidone and aripiprazole. However, the novelty of this finding is reduced by a careful search of the literature of the 1960s and 1970s where studies found that first-generation antipsychotics such as flupenthixol were prophylactic in bipolar disorder [22–25] as well as of course antimanic [26]. The early studies of antidepressants [27] found that chlorpromazine and other typical antipsychotics were equal to imipramine in depression. Thus, it is not clear if antibipolar prophylactic efficacy is unique to the new atypicals or has always been a property of dopamine D_2-blocking compounds. The sensitivity of bipolar patients to the weight gain and diabetogenic side effects of the new atypicals cannot be ignored. Many patients in our clinic are maintained on a combination of lithium and a typical antipsychotic such as perphenazine. However, others have done well on olanzapine, risperidone, ziprasidone and more recently quetiapine. Our impression with the antipsychotic medications has been similar to our experience with the anticonvulsant treatments: individual sensitivity and individual side effect acceptability seem to be more important than whether a particular compound is FDA approved for a particular indication. The advantage of long-term treatment of patients is that their individual optimal treatment can often be discovered in a manner that adds to evidence-based medicine. Evidence-based medicine in our clinic is no substitute for individualized care by a physician who knows the patient's baseline as well as the patient's response to medicine cessation that inevitably occurs at least several times over the long-term course of bipolar disorder.

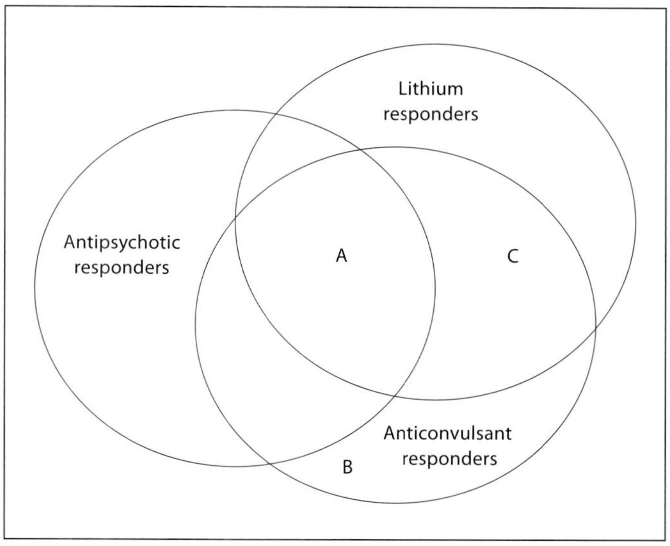

Fig. 1. Contemporary psychopharmacology of bipolar disorder. The enclosed areas represent 'sets' of patients successfully treatable prophylactically with each group of antibipolar drugs. A = Those who have done well with either lithium or an anticonvulsant or antipsychotics; B = those who have done well with an anticonvulsant but did not respond to antipsychotic or to lithium; C = those who achieve euthymia on lithium or an anticonvulsant but not on antipsychotics.

Bipolar depression is a source of less morbidity in our clinic than in reports from the USA or Europe, and we found that our patients, all bipolar I, have 3 manias for every depression [28]. This could be due to ethnic, genetic, or climactic factors or to the selection of patients, since our referrals come from hospital admissions that are almost always manic. Our clinical experience matches the recent American study [29] claiming that SSRIs are rarely useful in bipolar depression [30], but also the European experience that mixed noradrenergic-serotonergic antidepressants can alleviate bipolar depression [31]. We frequently see antidepressant-induced mania and try to treat mild-moderate bipolar depression with omega-3 fatty acids [32], folic acid [33], exercise [34, 35] and psychosocial intervention [36] to reduce mania risk.

The long-term follow-up in our clinic has led us to value Angst's distinction between Md (worse and more frequent manias than depressions), mD (worse and more frequent depressions than manias) and MD (equal depressions and manias) patients [37]. Antidepressants are clearly less risky in patients whose past manias have been

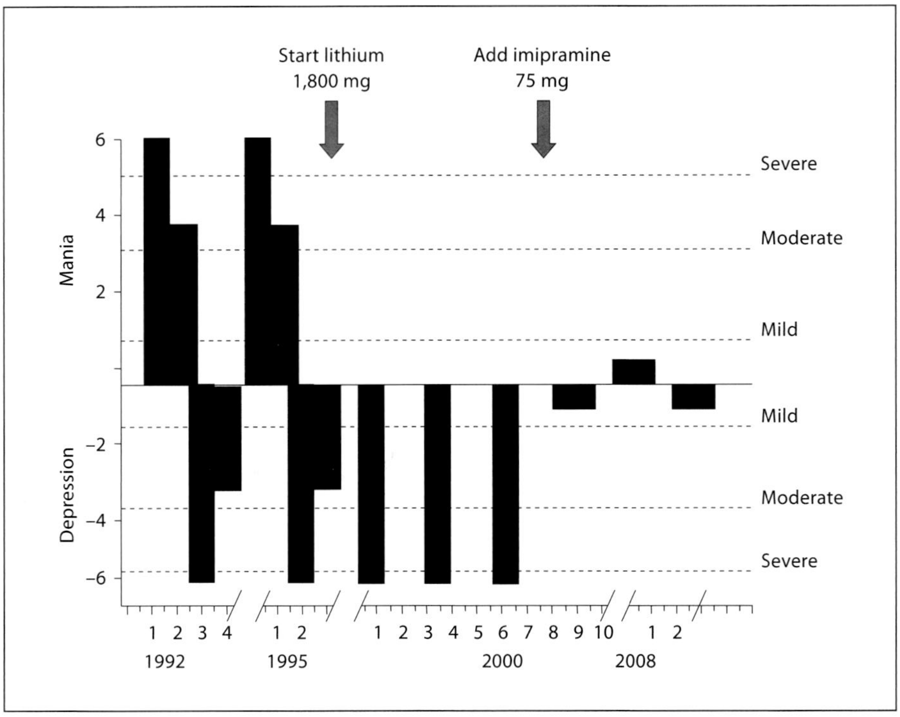

Fig. 2. Imipramine prophylaxis of residual depression in a lithium-treated bipolar patient. This patient's history is convincing to us of the utility of imipramine addition to lithium prophylaxis, and the lack of manic relapse in this patient. Longitudinal data are often more important for specific patients in our clinic than knowledge of the latest randomized trial of a large number of unselected patients.

mild than in those whose past manias were life-threatening. Some mD patients require chronic maintenance with mood stabilizers plus tricyclics or MAOI. As Prien et al. [38] and others have shown, such a strategy risks increased manic relapses in the general bipolar population but careful history and close acquaintance over time can minimize this risk. The cost-benefit ratio for specific patients is highly favorable (for instance, see the patient illustrated in figure 2).

Few patients in our clinic have participated in prophylactic controlled trials because our patients are well informed and have usually seen monotherapy versus placebo as too risky. Our patients have been more likely to participate, during relapse and inpatient interludes, in acute controlled trials of add-on drugs in mania [26, 39]. The largest outpatient prophylactic clinical study done in our clinic was an add-on study of phenytoin in addition to the patients' ongoing antibipolar regime [20].

Disability and Vocational Rehabilitation

The classic Kraepelinian diagnosis of bipolar disorder included the concept of restitutio in integrum, i.e. that patients were restored to their original normal personality and functioning between episodes. While this is true for perhaps a third of bipolar patients, long-term follow-up studies have shown that bipolar patients suffer from residual symptoms 30–50% of the time [40, 41], and that relapse occurs at a very high rate [40]. Up to two thirds of bipolar patients show significant decline in overall functioning (social adjustment and vocational activity) [42]. Total disability is not uncommon in bipolar disorder, and partial and/or temporary disability is considerably more common. A German study [43] found that half of 61 consecutive bipolar inpatients had been granted disability payments by their mid-forties; approximately 15% of the STEP-BD (Systematic Treatment Enhancement Program for Bipolar Disorder) sample reported their occupational status as 'disabled' [44]. Even when bipolar patients are virtually symptom free, their psychosocial functioning is still significantly below that of a healthy comparison group [45]. Divorce and marital separation are 2–3 times more likely than among the general population in the US [44, 46]. It is increasingly clear that the disorder is accompanied by neuropsychological [47] and cognitive [48] impairment even in the euthymic state. Euthymic bipolar patients show neuropsychological functioning which is superior to that of schizophrenic patients, but inferior to healthy comparison subjects [49]; problems are found most consistently in attention, memory, and executive function [50]. There is evidence that these cognitive deficits are related to poorer overall functional outcome, particularly the impairments in verbal memory and executive function [51], and that these deficits may be linked specifically to poorer occupational functioning [52].

This changing picture has influenced our approach to vocational rehabilitation, leading in many cases to a strategy of more modest goals. Thirty and even 20 years ago, we communicated to patients and families the expectation of full return to employment or studies within 3–6 months after hospital discharge from an acute affective episode. Now experience suggests an incremental approach to return to work, longer timetables for academic or vocational study programs, and more attention to risk factors such as irregular work hours or high-stress positions in the choosing of employment. We sometimes find ourselves having to interpret this message to families of patients, when their well-intentioned efforts to encourage the patient to do more and more are actually endangering the patient's well-being. While we continue to encourage patients to reach optimal levels of functioning in as many areas as possible, we have come to see that often, 'less is more'. More modest levels of achievement and functioning which can be sustained over time are preferable to overestimating abilities, over-reaching and exacerbating or even inducing recurrent affective episodes.

R.Z. is a 37-year-old single woman who is intelligent and artistic. She had little trouble finding work, but after performing successfully in a part-time capacity, she enlarged her workload to full time, and within a short time succumbed to a psychotic episode which entailed the loss of the job and an extended period of recovery. Only after she was able to see the pattern, and after receiving support for the idea that functioning well part-time was an acceptable outcome, did she limit her work hours and achieve a relatively long and stable remission. This also allowed the long delayed development of a satisfying romantic relationship. A second patient (R.W.), a social worker, experienced severe episodes while trying to retain various full-time, high-stress positions, but has now been able to remain healthy while working in her profession but at a part time low-stress position. A third patient (D.R.) suffered repeated exacerbations every time she attempted to return to work in her profession (nursing), but has been able to remain stable while working part time as a cosmetician. Patients have successfully completed academic degrees often only after accepting the need to spread the studies out over a longer than usual period of time, or accepting a less demanding course of study (such as a terminal MA, without thesis, as opposed to an MA with thesis).

Special emphasis is placed on setting and reaching one goal at a time, with adequate time between important milestones such as marriage, childbirth, commencing or completing studies, changing jobs, and moves to a new location.

Psychological Treatment

Bipolar disorder is clearly a biologically based illness with high heritability. A few decades ago, the expectation seemed to be that good blood levels of lithium alone would normalize the bipolar patient [53]. In early years, the lithium ion was thought to be a specific treatment for core bipolar disorder [54]. There is now considerable evidence that psychological or psychosocial interventions also have a significant impact on the course of the disease [55]. Research demonstrating the negative impact of high-expressed emotion in families on the relapse rates of schizophrenic patients [56] was soon found to be true also for affective disorders [57–59]. Several modes of psychotherapy have demonstrated usefulness as add-ons to treatment as usual in the management of bipolar disorder [60]. Interpersonal and social rhythm therapy focuses on personal problems such as grief, conflict resolution, and ineffective behaviors. It stresses the importance of maintaining regular schedules of sleep/wake, socializing, exercise, and work. This approach showed efficacy in a 2-year follow-up study [61]. Cognitive-behavioral therapy around medication compliance and depressive symptomology improved time to relapse, time well and social functioning over 1- and 2-year follow-up periods [62, 63]. See Miklowitz [36] for a comprehensive review of the evidence to date.

Psychoeducation with caregivers and family members of bipolar patients has also been shown to reduce risk of additional episodes [64]. We have made a concerted effort to include family members in the treatment of our clinic participants. Family members are involved in the ongoing follow-up treatment, and we endeavor to help them become active and informed members of the treatment team. A small booklet in Hebrew is given out to patients and families. This booklet, entitled *Bipolar Disorder (Manic-Depression)* [65], is one of a series which includes up-to-date information on a variety of medical and psychiatric disorders, presented in a well-organized and accessible manner, in clear everyday language. Additional recommended resources for English speakers range from a brief outline which succinctly summarizes main points describing the disorder and its management [66] to more comprehensive books. *The Bipolar Disorder Survival Guide* [66] is addressed directly to the bipolar patient, and conducts a sort of dialogue with the patient, addressing common fears, concerns and misunderstandings, imparting information together with a strong emphasis on self-help, self-care, and self-monitoring of mood states. The book includes many case vignettes, a section on additional resources and a complete list of references. A compre-

Table 1. Prescribing patterns in the Beer Sheva Lithium Clinic (1993 and 2008)

a Percentage of patients prescribed medications of each class

	Lithium	Anti-convulsant	Typical anti-psychotic	Atypical anti-psychotic	Anti-depressant
1993 (n = 54)	96	17	39	–	11
2008 (n = 155)	62	27	15	47	9

b Percentage of patients receiving monotherapy or polypharmacy (2, 3, or 4 medications)

	Mono-therapy	2 Medi-cations	3 Medi-cations	4 Medi-cations
1993 (n = 54)	46	44	7	2
2008 (n = 155)	48	40	11	1

hensive and user-friendly contribution to this literature is *Bipolar Disorder for Dummies* [67]. Deceptively lighthearted, with cartoons and the breezy contemporary style for which the 'For Dummies' series is noted, this book is actually a thorough and well-researched compendium of information and practical tips. The book is especially notable for its discussion of an 'emergency plan', a step-by-step guide to preparing for a possible crisis in the future. This book also includes a chapter on pediatric bipolar disorder, a section directed to friends and relatives who assist the patient, and a useful glossary of terms which can help improve patient-physician communication. Schou's [68] original booklet for patients and families is out of date.

We noted a special difficulty in working with parents of young adult bipolar patients. Whereas the patients, in many cases, had previously achieved a significant level of independence, the onset or exacerbation of the disorder resulted in a return to dependent or semi-dependent status – difficult for both patient and families to navigate. A support group for parents of adult bipolar patients was established, with the psychologist and a senior social worker as co-therapists. Over the course of several years, the group evolved from a more psychoeducational framework to one of mutual support; once the co-therapists were unable to continue, the group continued to meet, informally, without professional input.

A. is an unmarried man in his early 30s. Ill since his high school days but undiagnosed until his late 20s, A. was overly dependent on his parents and was seen as a failure by his younger sisters. In a series of family meetings utilizing reframing, structural interventions, goal setting and problem-solving techniques, expressed emotion on the part of the parents and the sisters was reduced and the patient was able to accept their appropriate help. He attained gradual improvement in functioning in a variety of roles ranging from housekeeping to gainful employment. The improvement in the family atmosphere facilitated the gradual movement of the patient towards a more appropriate level of independence, and for the first time, he was able to move out of the family home into his own apartment. Later, during three acute manic exacerbations, the high level of cooperation within the family and also between the family and the clinic staff enabled hospitalization to be avoided without endangering the patient or the family.

The Beer Sheva Mental Health Center Bipolar Disorder Clinic

About 10 years ago, the sign outside our clinic was changed from 'Lithium Clinic' to 'Bipolar Disorder Clinic' at around the time that only about half the patients were receiving lithium treatment (table 1). Each psychiatrist holds clinic 1 day per week from 9:00 a.m. to 3:00 p.m. Euthymic patients are generally invited to the clinic monthly; in rare emergencies, patients can be seen as often as several times per week. Patients are scheduled at 10- to 15-min intervals, and the delays which develop naturally over the day lead to an informal ongoing support group forming in the waiting area. The patient is seen by a psychiatrist *together with* the psychologist. During the appointment attention is given to overall assessment, to medication issues (dosage, side effects), and blood is drawn for levels of lithium or valproate or carbamazapine, or white blood cell count for clozapine patients, as required. The clinic visit is a prime opportunity for psychoeducation regarding the illness and its treatment. Topics include the nature of the disorder, symptoms, triggers and risk factors, medications and side effects, early detection of episodes, stress management, and lifestyle changes which can promote health and stability [69]. Family members often accompany the patient, and this is an important opportunity to recruit them to the treatment team, to provide psychoeducation and support, and to establish a rapport which can be critical in times of exacerbation. All patients, and their main supporters, are provided with the direct phone number of the psychologist. Scheduling and most routine inquiries are handled by the psychologist, who will contact the treating psychiatrist outside of clinic days only if necessary. The close involvement of the psychologist in the ongoing care of all patients generally allows the patients to feel that they have direct, continual access to a significant caregiver and at the same time demands on the physician's time are reduced.

The clinic appointment is a chance to hear about life events, stressors, conflicts, or other psychosocial problems which may require our intervention. Often time is insufficient and separate appointments are set up with the psychologist on nonclinic days. Some patients see the psychologist on a regular basis, for ongoing therapy; other patients and many families consult with her on an 'as needed' basis for interventions of varying length and intensity. The psychologist coordinates most rehabilitation programs and also prepares most entitlement forms for disability insurance, driving licenses and forensic evaluations. This strengthens the therapeutic alliance with the patient and enhances compliance with medication and clinic visits.

The Bipolar Clinic in its present form is conducive to both research and teaching opportunities. Clinic patients are well known to the staff and the average patient has been in treatment with the same staff for 10 years. Determinations of euthymic status or residual symptoms can be made with a high degree of confidence by comparison with a known baseline. Patients are used to blood being taken for determination of blood levels and very seldom object to small additional amounts of blood being drawn for research purposes (after obtaining written consent). Many important studies have resulted [70–72]. The clinic is often host to student groups (medical, nursing, and neuropsychology students) and provides a unique opportunity for students to see not the acute pathology of the inpatient wards but psychiatric patients who seem 'normal' and healthy despite their ongoing battle with a serious illness.

Table 2. Economics of the new lithium clinic model

Expenses per month – salaries	
Psychiatrist (R.H.B. and Y.B.), 1/5 position each, ×2 (2 psychiatrists, 1 day a week for each)	~10,000 NIS
Clinical psychologist (Y.O.), 3/4 position	~10,000 NIS
Total	~20,000 NIS
Income per month	
(from the Ministry of Health list of charges for services)	
Follow-up medication clinic visits – psychiatrist plus psychologist (charge for psychiatrist time only)	
160 visits/month × 94 NIS/visit	~15,000 NIS
Individual psychotherapy – psychologist	
24 visits/month × 200 NIS/visit	~5,000 NIS
Family and couples counseling – psychologist	
16 visits/month × 235 NIS/visit	~4,000 NIS
Psychosocial interventions and follow-up – psychologist	
40 visits/month × 108 NIS/visit	~4,000 NIS
Total	~28,000 NIS

The above model does not include laboratory costs, office space rent, computer and chart services and other overhead expenses. But it suggests that our model is within the guideline reimbursements projected for mental health services as part of the extension of National Health Insurance in Israel to include mental health on 'parity' basis with general medicine within the 4 national HMOs.

The Cost of the New Model for Bipolar Service Delivery

Pharmacological treatment of bipolar disorder is inadequate by itself; as Huxley and Baldessarini [42] conclude in their recent review, interpersonal, cognitive-behavioral, and psychoeducational interventions with bipolar patients can reduce risk of relapse, hospitalizations, and symptom severity as well as improve social and vocational outcomes. The addition of cognitive therapy to standard pharmacological treatment was found to be cost-effective as well as clinically superior [73], as the direct costs of the additional therapy were offset by reductions in other types of service usage. A collaborative care model utilizing a nurse care coordinator rather than a psychologist [74] has been shown to improve long-term clinical and functional outcomes in both a severely ill, frequently hospitalized cohort of bipolar patients [75] as well as in a less severely ill health maintenance organization sample [76]. The model was associated with a modest increase in cost as compared to treatment as usual for the HMO group [76]. For the more severely ill group, small increases in outpatient costs were more than offset by reductions in spending on inpatient care [75].

Under the proposed 'Reform' of Israeli mental health services, psychiatric hospitals and outpatient services will be reimbursed for services to members by the Health Maintenance Organizations which provide coverage to all citizens. Within the parameters of the current simulation figures in use, a very basic modeling of our clinic shows a favorable ratio of costs (salary) to income, as can be seen in table 2. Salary costs are computed based on total cost to employer, not only gross salary. Amounts shown are in New Israeli shekels (NIS); at current rates of exchange, 1 USD = 4 NIS (approximately).

Conclusion

Over the last 35 years, our Lithium Clinic → Bipolar Clinic has evolved. Bipolar patients often feel out of place in clinics designed for injectable or oral medication follow-up for schizophrenic patients. On the other hand, general psychiatric clinic practice may not provide specialized pharmacologic knowledge, availability of therapeutic drug level monitoring, rapid response to acute manic emergencies [30] and psychological services by a psychologist expert in mood swings. Changes over the last 35 years have not been only, or even primarily, exciting new pharmacologic treatments that help many but not all patients. Prognosis has been revised to include possible deterioration as well as drug nonresponse. Patients and families sometimes must hear this prognosis in an empathic way at an appropriate time. Psychoeducation and psychotherapy have been integrated in a way that might surprise early hopes for lithium as psychiatry's 'magic bullet' [77].

Acknowledgment

Thanks go to Eli Lepkifker who founded the first lithium clinic in Israel.

References

1 Schou M: Lithium in psychiatric therapy. Stock-taking after ten years. Psychopharmacologia 1959;1:65–78.
2 Fieve RR: The lithium clinic: a new model for the delivery of psychiatric services. Am J Psychiatry 1975;132:1018–1022.
3 Kerry RJ: A special lithium clinic. Dis Nerv Syst 1969;30:490–492.
4 Bey DR, Chapman RE, Tornquist KL: A lithium clinic. Am J Psychiatry 1972;129:468–470.
5 Ellenberg J, Salamon I, Meaney C: A lithium clinic in a community mental health center. Hosp Community Psychiatry 1980;31:834–836.
6 Schweitzer I, Davies B, Burrows G, Branton L, Turecek LR, Tiller J: The Royal Melbourne Hospital Lithium Clinic. Aust NZ J Psychiatry 1999;(33 Suppl):S35–S38.
7 Pietzcker A, Muller-Oerlinghausen B: The outpatient clinic for patients under chronic lithium or neuroleptic treatment as a phase-IV research tool. Pharmacopsychiatry 1984;17:162–167.
8 Coombs HI, Coombs RRH, Mee UG: Flame photometry; in Johnson FN (ed): Lithium Research and Therapy. London, Academic Press Inc, 1975, pp 167–172.
9 Cousins DA, Young AH: The armamentarium of treatments for bipolar disorder: a review of the literature. Int J Neuropsychopharmacol 2007;10:411–431.
10 Fountoulakis KN, Vieta E: Treatment of bipolar disorder: a systematic review of available data and clinical perspectives. Int J Neuropsychopharmacol 2008;11:999–1029.
11 Frye MA, Ketter TA, Leverich GS, Huggins T, Lantz C, Denicoff KD, Post RM: The increasing use of polypharmacotherapy for refractory mood disorders: 22 years of study. J Clin Psychiatry 2000;61:9–15.
12 Okuma T, Inanaga K, Otsuki S, Sarai K, Takahashi R, Hazama H, Mori A, Watanabe S: A preliminary double-blind study on the efficacy of carbamazepine in prophylaxis of manic-depressive illness. Psychopharmacology (Berl) 1981;73:95–96.
13 Post RM, Weiss SRB, Clark M, Chuang D-M, Hough C, Li H: Lithium, carbamazapine, and valproate in affective illness; in Manji HK, Bowden CL, Belmaker RH (eds): Bipolar Medications: Mechanisms of Action. Washington, American Psychiatric Press, Inc, 2000, pp 219–248.
14 Klein E, Bental E, Lerer B, Belmaker RH: Carbamazepine and haloperidol vs placebo and haloperidol in excited psychoses. A controlled study. Arch Gen Psychiatry 1984;41:165–170.
15 Emrich HM, von Zerssen D, Kissling W, Moller HJ: Therapeutic effect of valproate in mania. Am J Psychiatry 1981;138:256.
16 Bowden CL, Brugger AM, Swann AC, Calabrese JR, Janicak PG, Petty F, Dilsaver SC, Davis JM, Rush AJ, Small JG, et al: Efficacy of divalproex vs lithium and placebo in the treatment of mania. The Depakote Mania Study Group. JAMA 1994;271:918–924.
17 Bowden CL, Calabrese JR, Sachs G, Yatham LN, Asghar SA, Hompland M, Montgomery P, Earl N, Smoot TM, DeVeaugh-Geiss J: A placebo-controlled 18-month trial of lamotrigine and lithium maintenance treatment in recently manic or hypomanic patients with bipolar I disorder. Arch Gen Psychiatry 2003;60:392–400.
18 Levine J, Bersudsky Y, Nadri C, Yaroslavsky Y, Abed N, Mishory A, Agam G, Belmaker RH: Mechanism of action of new mood stabilizing drugs; in Soares JC, Gershon S (eds): Handbook of Medical Psychiatry. New York, Dekker, 2003, pp 793–806.
19 Prica C, Hascoet M, Bourin M: Antidepressant-like effect of lamotrigine is reversed by veratrine: a possible role of sodium channels in bipolar depression. Behav Brain Res 2008;191:49–54.
20 Mishory A, Winokur M, Bersudsky Y: Prophylactic effect of phenytoin in bipolar disorder: a controlled study. Bipolar Disord 2003;5:464–467.
21 Yatham LN, Kennedy SH, Schaffer A, Parikh SV, Beaulieu S, O'Donovan C, MacQueen G, McIntyre RS, Sharma V, Ravindran A, Young LT, Young AH, Alda M, Milev R, Vieta E, Calabrese JR, Berk M, Ha K, Kapczinski F: Canadian Network for Mood and Anxiety Treatments (CANMAT) and International Society for Bipolar Disorders (ISBD) collaborative update of CANMAT guidelines for the management of patients with bipolar disorder: update 2009. Bipolar Disord 2009;11:225–255.
22 Ahlfors UG, Baastrup PC, Dencker SJ, Elgen K, Lingjaerde O, Pedersen V, Schou M, Aaskoven O: Flupenthixol decanoate in recurrent manic-depressive illness. A comparison with lithium. Acta Psychiatr Scand 1981;64:226–237.
23 Bond DJ, Pratoomsri W, Yatham LN: Depot antipsychotic medications in bipolar disorder: a review of the literature. Acta Psychiatr Scand Suppl 2007;434:3–16.
24 Littlejohn R, Leslie F, Cookson J: Depot antipsychotics in the prophylaxis of bipolar affective disorder. Br J Psychiatry 1994;165:827–829.
25 Tohen M, Zhang F, Taylor CC, Burns P, Zarate C, Sanger T, Tollefson G: A meta-analysis of the use of typical antipsychotic agents in bipolar disorder. J Affect Disord 2001;65:85–93.

26 Biederman J, Lerner Y, Belmaker RH: Combination of lithium carbonate and haloperidol in schizo-affective disorder: a controlled study. Arch Gen Psychiatry 1979;36:327–333.
27 Fink M, Klein DF, Kramer JC: Clinical efficacy of chlorpromazine-procyclidine combination, imipramine and placebo in depressive disorders. Psychopharmacologia 1965;7:27–36.
28 Osher Y, Yaroslavsky Y, el-Rom R, Belmaker RH: Predominant polarity of bipolar patients in Israel. World J Biol Psychiatry 2000;1:187–189.
29 Sachs GS, Nierenberg AA, Calabrese JR, Marangell LB, Wisniewski SR, Gyulai L, Friedman ES, Bowden CL, Fossey MD, Ostacher MJ, Ketter TA, Patel J, Hauser P, Rapport D, Martinez JM, Allen MH, Miklowitz DJ, Otto MW, Dennehy EB, Thase ME: Effectiveness of adjunctive antidepressant treatment for bipolar depression. N Engl J Med 2007;356:1711–1722.
30 Belmaker RH: Bipolar disorder. N Engl J Med 2004;351:476–486.
31 Gijsman HJ, Geddes JR, Rendell JM, Nolen WA, Goodwin GM: Antidepressants for bipolar depression: a systematic review of randomized, controlled trials. Am J Psychiatry 2004;161:1537–1547.
32 Nemets B, Stahl Z, Belmaker RH: Addition of omega-3 fatty acid to maintenance medication treatment for recurrent unipolar depressive disorder. Am J Psychiatry 2002;159:477–479.
33 Coppen A, Bolander-Gouaille C: Treatment of depression: time to consider folic acid and vitamin B_{12}. J Psychopharmacol 2005;19:59–65.
34 Mead GE, Morley W, Campbell P, Greig CA, McMurdo M, Lawlor DA: Exercise for depression. Cochrane Database Syst Rev 2008;4:CD004366.
35 Morgan AJ, Jorm AF: Self-help interventions for depressive disorders and depressive symptoms: a systematic review. Ann Gen Psychiatry 2008;7:13.
36 Miklowitz DJ: Adjunctive psychotherapy for bipolar disorder: state of the evidence. Am J Psychiatry 2008;165:1408–1419.
37 Angst J: Clinical typology of bipolar illness; in Belmaker RH, van Praag HM (eds): Mania: An Evolving Concept. New York, Spectrum Publications, Inc, 1980, pp 61–76.
38 Prien RF, Kupfer DJ, Mansky PA, Small JG, Tuason VB, Voss CB, Johnson WE: Drug therapy in the prevention of recurrences in unipolar and bipolar affective disorders. Report of the NIMH Collaborative Study Group comparing lithium carbonate, imipramine, and a lithium carbonate-imipramine combination. Arch Gen Psychiatry 1984;41:1096–1104.
39 Mishory A, Yaroslavsky Y, Bersudsky Y, Belmaker RH: Phenytoin as an antimanic anticonvulsant: a controlled study. Am J Psychiatry 2000;157:463–465.
40 Judd LL, Akiskal HS, Schettler PJ, Endicott J, Maser J, Solomon DA, Leon AC, Rice JA, Keller MB: The long-term natural history of the weekly symptomatic status of bipolar I disorder. Arch Gen Psychiatry 2002;59:530–537.
41 Judd LL, Schettler PJ, Solomon DA, Maser JD, Coryell W, Endicott J, Akiskal HS: Psychosocial disability and work role function compared across the long-term course of bipolar I, bipolar II and unipolar major depressive disorders. J Affect Disord 2008;108:49–58.
42 Huxley N, Baldessarini RJ: Disability and its treatment in bipolar disorder patients. Bipolar Disord 2007;9:183–196.
43 Brieger P, Bloink R, Rottig S, Marneros A: Disability payments due to unipolar depressive and bipolar affective disorders. Psychiatr Prax 2004;31:203–206.
44 Kogan JN, Otto MW, Bauer MS, Dennehy EB, Miklowitz DJ, Zhang HW, Ketter T, Rudorfer MV, Wisniewski SR, Thase ME, Calabrese J, Sachs GS: Demographic and diagnostic characteristics of the first 1,000 patients enrolled in the Systematic Treatment Enhancement Program for Bipolar Disorder (STEP-BD). Bipolar Disord 2004;6:460–469.
45 Judd LL, Akiskal HS, Schettler PJ, Endicott J, Leon AC, Solomon DA, Coryell W, Maser JD, Keller MB: Psychosocial disability in the course of bipolar I and II disorders: a prospective, comparative, longitudinal study. Arch Gen Psychiatry 2005;62:1322–1330.
46 Suppes T, Leverich GS, Keck PE, Nolen WA, Denicoff KD, Altshuler LL, McElroy SL, Rush AJ, Kupka R, Frye MA, Bickel M, Post RM: The Stanley Foundation Bipolar Treatment Outcome Network. 2. Demographics and illness characteristics of the first 261 patients. J Affect Disord 2001;67:45–59.
47 Savitz J, Solms M, Ramesar R: Neuropsychological dysfunction in bipolar affective disorder: a critical opinion. Bipolar Disord 2005;7:216–235.
48 Mur M, Portella MJ, Martinez-Aran A, Pifarre J, Vieta E: Long-term stability of cognitive impairment in bipolar disorder: a 2-year follow-up study of lithium-treated euthymic bipolar patients. J Clin Psychiatry 2008;69:712–719.
49 Goodwin GM, Martinez-Aran A, Glahn DC, Vieta E: Cognitive impairment in bipolar disorder: neurodevelopment or neurodegeneration? An ECNP expert meeting report. Eur Neuropsychopharmacol 2008;18:787–793.
50 Torres IJ, Boudreau VG, Yatham LN: Neuropsychological functioning in euthymic bipolar disorder: a meta-analysis. Acta Psychiatr Scand Suppl 2007;434:17–26.
51 Martinez-Aran A, Vieta E, Torrent C, Sanchez-Moreno J, Goikolea JM, Salamero M, Malhi GS, Gonzalez-Pinto A, Daban C, Alvarez-Grandi S, Fountoulakis K, Kaprinis G, Tabares-Seisdedos R, Ayuso-Mateos JL: Functional outcome in bipolar disorder: the role of clinical and cognitive factors. Bipolar Disord 2007;9:103–113.
52 Dickerson FB, Boronow JJ, Stallings CR, Origoni AE, Cole S, Yolken RH: Association between cognitive functioning and employment status of persons with bipolar disorder. Psychiatr Serv 2004;55:54–58.
53 Solomon DA, Ristow WR, Keller MB, Kane JM, Gelenberg AJ, Rosenbaum JF, Warshaw MG: Serum lithium levels and psychosocial function in patients with bipolar I disorder. Am J Psychiatry 1996;153:1301–1307.
54 Soares JC, Gershon S: The lithium ion: a foundation for psychopharmacological specificity. Neuropsychopharmacology 1998;19:167–182.
55 Beynon S, Soares-Weiser K, Woolacott N, Duffy S, Geddes JR: Psychosocial interventions for the prevention of relapse in bipolar disorder: systematic review of controlled trials. Br J Psychiatry 2008;192:5–11.
56 Vaughn CE, Leff JP: The influence of family and social factors on the course of psychiatric illness. A comparison of schizophrenic and depressed neurotic patients. Br J Psychiatry 1976;129:125–137.
57 Leff J, Vaughn C: The interaction of life events and relatives' expressed emotion in schizophrenia and depressive neurosis. Br J Psychiatry 1980;136:146–153.
58 Morris CD, Miklowitz DJ, Waxmonsky JA: Family-focused treatment for bipolar disorder in adults and youth. J Clin Psychol 2007;63:433–445.
59 Priebe S, Wildgrube C, Muller-Oerlinghausen B: Lithium prophylaxis and expressed emotion. Br J Psychiatry 1989;154:396–399.
60 Fountoulakis KN, Vieta E, Siamouli M, Valenti M, Magiria S, Oral T, Fresno D, Giannakopoulos P, Kaprinis GS: Treatment of bipolar disorder: a complex treatment for a multi-faceted disorder. Ann Gen Psychiatry 2007;6:27.
61 Frank E, Kupfer DJ, Thase ME, Mallinger AG, Swartz HA, Fagiolini AM, Grochocinski V, Houck P, Scott J, Thompson W, Monk T: Two-year outcomes for interpersonal and social rhythm therapy in individuals with bipolar I disorder. Arch Gen Psychiatry 2005;62:996–1004.
62 Lam DH, Hayward P, Watkins ER, Wright K, Sham P: Relapse prevention in patients with bipolar disorder: cognitive therapy outcome after 2 years. Am J Psychiatry 2005;162:324–329.
63 Lam DH, Watkins ER, Hayward P, Bright J, Wright K, Kerr N, Parr-Davis G, Sham P: A randomized controlled study of cognitive therapy for relapse prevention for bipolar affective disorder: outcome of the first year. Arch Gen Psychiatry 2003;60:145–152.

64 Reinares M, Colom F, Sanchez-Moreno J, Torrent C, Martinez-Aran A, Comes M, Goikolea JM, Benabarre A, Salamero M, Vieta E: Impact of caregiver group psychoeducation on the course and outcome of bipolar patients in remission: a randomized controlled trial. Bipolar Disord 2008;10:511–519.
65 Weitzman A, Tal V: Bipolar Disorder (Manic-Depression). Rosh Haayin, Prolog Publishing House, 2001.
66 Miklowitz DJ: The Bipolar Disorder Survival Guide. New York, Guilford Press, 2002.
67 Fink DF, Kraynak J: Bipolar Disorder for Dummies. New Jersey, Wiley Publishing Inc, 2005.
68 Schou M: Lithium Treatment of Manic-Depressive Illness, ed 3. Basel, Karger, 1986.
69 Colom F, Vieta E: Psychoeducation Manual for Bipolar Disorder. Cambridge, Cambridge University Press, 2006.
70 Belmaker RH, Shapiro J, Vainer E, Nemanov L, Ebstein RP, Agam G: Reduced inositol content in lymphocyte-derived cell lines from bipolar patients. Bipolar Disord 2002; 4:67–69.
71 Osher Y, Bersudsky Y, Silver H, Sela BA, Belmaker RH: Neuropsychological correlates of homocysteine levels in euthymic bipolar patients. J Affect Disord 2008;105:229–233.
72 Shamir A, Ebstein RP, Nemanov L, Zohar A, Belmaker RH, Agam G: Inositol monophosphatase in immortalized lymphoblastoid cell lines indicates susceptibility to bipolar disorder and response to lithium therapy. Mol Psychiatry 1998;3:481–482.
73 Lam DH, McCrone P, Wright K, Kerr N: Cost-effectiveness of relapse-prevention cognitive therapy for bipolar disorder: 30-month study. Br J Psychiatry 2005;186:500–506.
74 Bauer MS, McBride L, Williford WO, Glick H, Kinosian B, Altshuler L, Beresford T, Kilbourne AM, Sajatovic M: Collaborative care for bipolar disorder. 1. Intervention and implementation in a randomized effectiveness trial. Psychiatr Serv 2006;57:927–936.
75 Bauer MS, McBride L, Williford WO, Glick H, Kinosian B, Altshuler L, Beresford T, Kilbourne AM, Sajatovic M: Collaborative care for bipolar disorder. 2. Impact on clinical outcome, function, and costs. Psychiatr Serv 2006;57:937–945.
76 Simon GE, Ludman EJ, Bauer MS, Unutzer J, Operskalski B: Long-term effectiveness and cost of a systematic care program for bipolar disorder. Arch Gen Psychiatry 2006;63:500–508.
77 Gershon S, Angrist B, Shopsin B: Pharmacological agents as tools in psychiatric research; in Gershon ES, Belmaker RH, Kety SS, Rosenbaum M (eds): The Impact of Biology on Modern Psychiatry. New York, Plenum Press, 1977.

Neuropsychobiology

Neuropsychobiology 2010;62:27–35
DOI: 10.1159/000314307

Published online: May 7, 2010

Lithium: Still a Cornerstone in the Long-Term Treatment in Bipolar Disorder?

Alessandra M.A. Nivoli[a, b] Andrea Murru[a] Eduard Vieta[a]

[a]Bipolar Disorder Program, IDIBAPS, CIBERSAM, Clinical Institute of Neuroscience, Hospital Clinic, University of Barcelona, Barcelona, Spain; [b]Clinical Institute of Psychiatry, Department of Neuroscience, University of Sassari, Sassari, Italy

Key Words
Bipolar disorder, long-term treatment · Lithium · Lithium prophylaxis

Abstract
Background: Scientific literature considers lithium a key treatment for the acute and long-term management of bipolar disorder (BD). Despite its worldwide clinical use, the effectiveness of lithium has been questioned. The aim of this work is to critically review the available data on randomized controlled trials (RCTs) concerning long-term lithium treatment. ***Methods:*** A systematic search for long-term treatment RCTs with at least 6 months of follow-up was performed. Six RCTs enrolling 1,561 bipolar I and II patients of adult and pediatric age, randomizing 534 to lithium, were identified. All studies are controlled trials sponsored by industry, investigating new treatments for BD, with lithium as an active comparator, and therefore not specifically designed to study lithium efficacy or safety. ***Results:*** Results from earliest studies suggest a high effectiveness of lithium against both mania and depression, while more recent studies highlight lithium as more effective than placebo in mania and hypomania, without significant evidence in depression. Lithium does not achieve significant differences in efficacy when compared with divalproex; it seems less effective than lamotrigine in preventing depression and less effective than olanzapine in manic and mixed episodes. ***Conclusions:*** Despite a number of methodological issues (enriched designs, unbalanced samples, potential inclusion of lithium nonresponders in some studies), lithium appears to have a clear antimanic prophylactic activity and some efficacy in the prevention of depression. Lithium should still have a major role in the long-term treatment of BD.

Introduction

From a historical point of view, lithium has been well known since the end of the 19th century as treatment for a number of medical conditions. Discovered by Johan Arfwedson in 1817, lithium was used for the first time in medicine in 1843 by Alexander Ure [1]. Lithium's effects on manic-depressive illness were discovered by chance in 1949 by John Cade who observed its sedating effect [2, 3]. Mogens Schou established the effectiveness of lithium for the treatment of bipolar disorder (BD) [4]. The Food and Drug Administration approved lithium in 1971 for treating manic episodes of manic-depressive illness.

The mechanisms of action of lithium are still unclear and the biochemical basis for its therapeutic effects is not exactly known [5, 6]. It acts in multiple sites inside the neuron as an inhibitor of key enzymes involving multiple

KARGER

Fax +41 61 306 12 34
E-Mail karger@karger.ch
www.karger.com

0302–282X/10/0621–0027$26.00/0

Accessible online at:
www.karger.com/nps

Eduard Vieta, MD, PhD
Neuroscience Institute, Hospital Clinic, University of Barcelona, IDIBAPS
Villaroel 170
ES–08036 Barcelona (Spain)
Tel. +34 93 227 5401, Fax +34 93 227 5477, E-Mail evieta@clinic.ub.es

neurotransmitter systems [7, 8], inducing a number of effects at presynaptic, postsynaptic and transcriptional levels, both with short-term effects (manifested soon after lithium's application and mediated through the available cellular apparatus) and long-term effects (presumably based on selective changes in gene expression and appearing after several weeks) [9–12].

Lithium is used as a first-line treatment for BD, both in the acute manic phase and in prophylaxis, for recurrent manic and depressive episodes. International guidelines consider lithium as a key treatment for the management of BD and recommend it in monotherapy as the first-line long-term treatment [13–16]. Individual episodes of depression and mania in the context of a BD can usually be successfully treated, but relapses are common and their prevention is therefore important to reduce the burden of BD, one of the leading causes of disability worldwide, which carries a substantial risk of suicide [17]. Lithium was the standard treatment for relapse prevention over several decades since the publication of early trials on BD treatment in the 1960s and 1970s. The latest studies, methodologically defined as randomized controlled trials (RCTs), have been designed in order to study the efficacy and safety of new drugs and they have used lithium as a comparator, in monotherapy or in association, producing substantial new data on its long-term efficacy.

Despite its worldwide clinical use, the effectiveness of lithium in improving the course of illness, especially in the long-term, has been questioned [18]. The challenge still remains to decide how and when it can best be employed, alone or in combination with other effective agents, to improve the outcome of BD in clinical practice [19]. In the last few years, new RCTs have contributed to strengthen data on lithium efficacy and tolerability. The aim of our work is to systematically review these most recent, available data on RCTs concerning long-term treatment of BD in which lithium has been evaluated. In our review, we chose to include RCTs with and without placebo arms, considering both efficacy and tolerability outcomes, in order to summarize the state of the art concerning methodologically rigorous data.

Methods

Search Strategy

A systematic search for RCTs concerning long-term lithium treatment was performed using MEDLINE and Cochrane Central Register of Controlled Trials (CENTRAL), considering a time period beginning on January 1, 2000. We chose to include findings from 2000 on in order to take into account the studies with the most modern designs. The search was completed on January 15, 2009. A cross-check between obtained references was done.

In order to locate RCTs, the search was performed using the key words 'maintenance bipolar disorder' alone and combined with the following strings: 'lithium', 'lithium RCT', 'lithium placebo controlled', 'RCT', 'antiepileptic', 'atypical antipsychotic', 'amisulpride', 'aripiprazole', 'carbamazepine', 'clozapine', 'gabapentin', 'haloperidol', 'lamotrigine', 'olanzapine', 'oxcarbazepine', 'perphenazine', 'quetiapine', 'risperidone', 'typical antipsychotic', 'valproate' ('divalproex', 'valproic acid'), 'ziprasidone'. The authors searched for placebo-controlled as well as for clinical trials with an active comparator, with lithium used as monotherapy.

Moreover, web pages containing lists of clinical trials were also scanned. These sites included: http://clinicaltrials.gov and http://www.clinicalstudyresults.org. An additional search included previous systematic reviews in order to trace trials published before 2000. Data extraction and quality assessment were carried out by one reviewer and checked independently by a second reviewer.

Inclusion/Exclusion Criteria

We included in this review all RCTs with at least 6 months of follow-up that investigated the effectiveness and/or tolerability of lithium used as maintenance treatment for BD. Only articles in English having undergone peer review were considered in the present report. Studies which included both unipolar and bipolar patients were included only if the data for bipolar patients could be extracted separately. Non-RCT studies were excluded from this review, as well as studies with mixed samples, add-on studies and other studies with lithium in combination. Duplicated publications of trials were identified and only data from a single study were included in the review.

Data Extraction Strategy

The first step was to decide, on the basis of the abstract or full text, whether or not the article was an RCT. The paper should deal with long-term lithium therapy in BD and report study design, sample description, outcome measures, and adverse events. Each citation was individually reviewed to identify all prospective randomized studies in which lithium prophylaxis as monotherapy was given for at least 6 months. Data were classified into three groups: (1) data of interest, (2) duplicated data and (3) data of no interest, according to the previously described criteria. RCTs were then classified into those concerning lithium monotherapy with and without a placebo arm. Where available, data for suicide were analyzed for each comparison. The primary outcome measure was time to any mood episode. Secondary outcome measures were manic and depressive relapses, and change from baseline in the score of rating scales.

Results

Results of the Literature Search

The search returned 832 papers from MEDLINE and 155 papers from CENTRAL. The great majority of them had to be excluded as they were opinions of review papers, observational and open studies; some of the RCTs

Table 1. Designs of the studies considered

References	Study design	Sponsored drug	Primary aim	Arms	Duration/ open label	Enriched design	Li vs. PBO	Li levels mEq/l
Bowden et al. [20] 2000	randomized, double-blind, parallel-group, multicenter study	DVP	comparison of efficacy and safety of DVP vs. Li and vs. PBO	2:1:1 DVP:Li:PBO	12 months/ 12 weeks	DVP, Li, DVP + Li, PBO	N	0.8–1.2
Bowden et al. [21] 2003	randomized, double-blind, placebo-controlled, parallel-group, multicenter study	LMT	comparison of efficacy and tolerability of LMT and Li vs. PBO	1:1:1 LMT:Li:PBO	18 months/ 8–16 weeks	LMT	Y	0.8–1.1
Calabrese et al. [22] 2003	randomized, double-blind, parallel-group, placebo-controlled, multicenter study	LMT	comparison of efficacy and tolerability of LMT and Li vs. PBO	3:1:1 LMT:Li:PBO	18 months/ 8–16 weeks	LMT	Y	0.92
Calabrese et al. [23] 2005	randomized, double-blind, parallel-group study	DVP	superiority of DVP vs. Li	1:1 DVP:Li	20 months/ 12–20 weeks	DVP, Li	–	0.8 ± 1.1
Tohen et al. [25] 2005	randomized, double-blind, controlled study	OLZ	OLZ's noninferiority to Li	1:1 OLZ:Li	12 months/ 6–12 weeks	OLZ, Li	–	0.8 ± 1.4
Findling et al. [24] 2005	randomized, double-blind, multiphase, single-site study	DVP	superiority of efficacy and safety of DVP vs. Li	1:1 Li:DVP	18 months/ 20 weeks	DVP, Li	–	0.6–1.2

Li = Lithium.

were on a complex combination or adjunctive therapy and many focused on different agents. At last, only a small minority of papers fulfilled our inclusion criteria and could be considered as RCTs on lithium monotherapy in maintenance treatment for BD. This final selection left us with 6 papers: 3 with placebo arms [20–22], and 3 without a placebo arm [23–25]. Included articles are summarized in table 1; 5 reviews were evaluated [26–30].

RCTs before 2000

The results of controlled trials with a placebo arm in the early 70s globally showed the effectiveness of lithium in the prevention of any mood episode in bipolar patients, but the generally small size of the samples studied does not permit a more formal analysis [29]. Placebo-treated patients showed a threefold increased rate of manic relapses and a twofold increased rate of depressive relapses compared to lithium-maintained patients: within 0.5–2 years of observation, the percentage of relapse was 31% in the lithium group and 76% in the placebo group [27].

Studies without a placebo arm carried out before 2000 showed a trend for lithium to perform slightly (but not significantly) better than carbamazepine. Low doses of lithium did not differ from placebo in a comparison of cognitive behavior treatment [29].

RCTs after 2000

The analyzed RCTs on long-term lithium monotherapy in BD included recent randomized trials sponsored by industry, investigating new treatment for BD that integrated lithium as a comparator, not designed to have lithium efficacy or safety as a primary aim. Six RCTs enrolling 1,561 bipolar I and II patients, randomizing 534 to lithium, were identified, with and without a placebo comparison arm.

The selected trials with a placebo arm included 3 randomized, double-blind, placebo-controlled, parallel-group, multicenter studies: 1 considers divalproex (DVP) [20] and 2 consider lamotrigine (LMT) [21, 22]. All these studies evaluate lithium as a comparator with efficacy and/or tolerability of the studied drug being the primary aim. Samples all include bipolar I patients, following the DSM-IV criteria, with only 1 study using DSM-III criteria [20]. Calabrese et al. [23] and Findling et al. [24] also included bipolar II patients. Findings from studies are reported below.

In a 12-month placebo-controlled study which compares the efficacy of DVP with lithium and placebo as prophylactic therapy, lithium proved to be more effective than placebo, not showing any significant difference compared to DVP in patients with mild forms of BD [20]. Drug efficacy is defined considering time to intervention, either by a syndromic (time to any mood episode by clinical judgment) or by a symptomatic point of view, evaluated through rating scales such as the Mania Rating Scale and Depressive Syndrome Scale. The authors enrolled 571 bipolar I patients in the open-label phase, manic/partially or totally recovered from a manic episode but not depressed, and randomized 372 patients to double-blind maintenance treatment within 3 months of the onset of a manic/hypomanic episode (Mania Rating Scale total score <11; Depressive Syndrome Scale total score <13; Global Assessment of Functioning Scale score >60). Patients treated with DVP continued treatment significantly longer than those with lithium (198 ± 152 days vs. 152 ± 148 days; $p = 0.02$). Time to any mood episode did not differ significantly in the three groups, as well as mean change on RS total scores. The median times to 50% survival without any mood episode were 40 weeks for DVP, 24 weeks for lithium and 28 weeks for placebo. Lithium has been found to show more side effects than placebo, such as nausea ($p = 0.05$), diarrhea ($p = 0.02$) and tremor ($p < 0.001$), increases in white blood cell count, uric acid, calcium and creatinine levels; when compared to DVP, it showed less sedation, infections and tinnitus, and increased polyuria and polydipsia ($p = 0.01$), tachycardia ($p = 0.04$), akathisia ($p = 0.04$), and xerophthalmia ($p = 0.03$).

In the 18-month placebo-controlled study on LMT by Bowden et al. [21], lithium was found to be more effective compared to placebo in preventing any mood episode ($p = 0.003$), manic/hypomanic episodes ($p = 0.006$) but not depressive episodes ($p = 0.17$) in a sample of 175 currently or recently manic/hypomanic bipolar I patients. Depressive episodes constituted the majority of events (56%) requiring intervention for the lithium group. No significant differences between lithium and LMT in time to intervention are referred on any mood episode ($p = 0.46$). In the overall survival analysis, lithium did not show any difference compared to placebo. However, in the lithium group, a significant decrease in symptoms on the Mania Rating Scale was observed, compared to the LMT group ($p = 0.03$). Assessment of safety outcomes showed that lithium significantly causes diarrhea compared to both LMT ($p = 0.002$) and placebo ($p = 0.009$), while for LMT the authors report a significantly higher prevalence of headache ($p = 0.02$).

Calabrese et al. [22], in their 18-month, placebo-controlled, double-blind, parallel-group study conducted on a sample of 463 recently depressed bipolar I patients, highlighted similar findings comparing lithium to placebo and LMT, with lithium predominantly effective against mania. Lithium was statistically superior to placebo in prolonging time to intervention for any mood episode ($p = 0.03$); however, the authors did not find any difference between lithium and LMT ($p = 0.9$). Median times to treatment intervention were 93 days for placebo, 170 days for lithium and 200 days for LMT. Time to intervention for manic or hypomanic episodes was significantly longer for lithium than for placebo ($p = 0.026$). Concerning changes in symptom severity, rating scales indicate differences in depressive symptoms, evaluated with the Hamilton Depression Rating Scale. The total score showed a significant decrease in symptoms ($p < 0.05$) for lithium compared to placebo; the same can be observed in the Global Assessment Scale and Clinical Global Impression-Severity of Illness Scale. Side effects recorded in the lithium group were similar to a previous study: significantly more somnolence and tremor ($p < 0.05$). Lithium shows more diarrhea and tremor when compared to LMT ($p < 0.005$).

The selected trials without a placebo arm included 3 studies published in 2005 comparing lithium with DVP [23, 24] and with olanzapine (OLZ) [25]. They are all double-blind, randomized, controlled studies including bipolar I and II adult patients, with the study by Findling et al. [24] focusing on pediatric ones.

Calabrese et al. [23] carried out a DVP 20-month study in 60 recently hypomanic/manic rapid cyclic bipolar I and II patients, testing the hypothesis that DVP would be more effective than lithium. They found no support for their thesis as lithium did not significantly differ from DVP in preventing any mood episode ($p > 0.05$) as well as in improving symptom severity on the Hamilton Depression Rating Scale and Young Mania Rating Scale. The incidence of tremor ($p < 0.02$), polyuria and polydipsia ($p = 0.001$) in the lithium group was found to be higher than in the DVP group, but premature discontinuation due to adverse events did not differ between treatment arms.

Tohen et al. [25], in their 12-month study, evaluated OLZ's noninferiority to lithium in the maintenance treatment of 431 recently manic/hypomanic bipolar I patients. The authors highlighted the noninferiority of OLZ relative to lithium in preventing any mood episodes. Considering pole-specific recurrences, OLZ and lithium did not differ significantly in the proportion of patients with a depressive recurrence. Side effects such as insomnia and nausea have been recorded more frequently in the lithium group ($p < 0.001$ and $p = 0.02$, respectively) as well as worsening of mania ($p < 0.001$). Depression NAS ($p = 0.01$), hypersomnia ($p = 0.03$), 7% or more weight gain and increase in cholesterol levels of nonclinical interest

($p < 0.001$) have been recorded more frequently in the OLZ-treated patients.

A pediatric sample of 60 bipolar I and II patients was evaluated for 18 months by Findling et al. [24], who found out that lithium and DVP maintenance treatments did not differ in survival time until emerging symptoms of relapse ($p = 0.55$) or survival time until discontinuation for any reason ($p = 0.72$). Lithium more frequently produced emesis ($p = 0.05$), enuresis ($p = 0.002$) and polydipsia ($p = 0.05$).

Discussion

Earliest placebo-controlled studies suggest a high effectiveness of lithium in the prophylaxis of both mania and depression in BD [31–35]. Results from more recent studies on novel treatments including a lithium comparison group permit to highlight, so far, that lithium is more effective than placebo in preventing new episodes of mood disturbance, as confirmed by previously cited literature.

Efficacy Findings

Comparison of lithium with placebo was not performed in the first study of Bowden et al. [20], despite the presence of a placebo arm. On the other hand, the following studies compared lithium with placebo, even if it was not their primary aim [21, 22]. Lithium was found to be superior to placebo in preventing mania and hypomania, both in recently manic/hypomanic [21] and depressed [22] bipolar I patients, but did not show significant evidence in preventing depression. Data on comparisons between lithium and DVP did not produce significant differences in time to any mood episode in the first DVP study [20], and in the others, DVP RCTs [23, 24] failed to demonstrate the superiority of DVP over lithium. Lithium did not differ from LMT in the efficacy of preventing any mood episode, while a trend in prolonging time to manic/hypomanic and mixed episodes favored lithium over LMT [21]. In the following comparison with recently depressed bipolar patients, lithium significantly delayed time to treatment for episodes of mania, but not depression, and it did not exhibit evidence of worsening of any phase of the illness. These findings have to be carefully interpreted because LMT studies are not designed for a direct comparison between the two treatment arms. OLZ and lithium did not statistically differ in preventing any mood episode recurrence according to symptomatic rating scale criteria. The authors state the superiority of OLZ over lithium in preventing recurrence of manic and mixed episodes, but not depression.

Methodological Issues

Published RCTs on long-term lithium monotherapy in BD are not easy to interpret and several methodological issues have to be analyzed: study design, presence of a placebo arm, sponsored drug, sample characteristics (inclusion/exclusion criteria, size and statistical power), primary and secondary outcome definitions and safety outcomes. In the following paragraphs, we will try to give a critical overview over these issues.

First, the study design limits extraction of data on lithium efficacy and safety: all the included studies are meant to assess drugs (sponsored drug) other than lithium, which is not the target of efficacy and safety proposed outcomes. Moreover, comparison between lithium and other drugs is problematic because of the unbalanced design of placebo-controlled studies and lithium groups are not always adequately powered. Among placebo-controlled trials, only Bowden's DVP study was specifically designed to compare novel treatment to lithium. Enriched designs during the open-label phase are evaluated in all examined long-term studies for lithium and novel treatment, with LMT studies being enriched for sponsored drug only. This may have affected the outcome values, by selecting patients showing better response and/or better tolerability to novel treatment [36]. The inclusion of a placebo arm in the study design is important as well, in order to provide sufficiently rigorous evidence [37]. Three out of the 6 studies include a placebo arm, 2 of them designed to perform lithium-placebo direct comparison. A somewhat surprising result was found in Bowden's DVP study. The relatively good outcomes of placebo have to be read cautiously because of the study's very design. Moreover, mild patients were included in the sample, while more severely ill patients may have been reluctant to enroll for the risk of the 12-month placebo assumption, and drug-placebo differences are known to be greater when patients with more severe forms of illness are studied [38]. All these aspects may have contributed to the surprisingly good outcomes of the placebo group in this study.

Sample selection needs to be discussed, too. Inclusion criteria in the different studies varied (table 2), making comparison somehow difficult. Diagnostic criteria usually follow DSM-IV-TR, with the exception of Bowden's DVP study (DSM-III). Bipolar I patients are the most represented diagnostic group in all studies; bipolar II and rapid cyclers constitute diagnostic subtypes which seem

Table 2. Sample characteristics of the studies considered in the present review

References	Patient selection		Sample		
	diagnosis/clinical criteria	rating scales	general size	Li arm size	power for Li
Bowden et al. [20] 2000	BPI DSM-III-R (M/Eu)	MRS <11, DSS <13, GAS >60	372	91	Y
Bowden et al. [21] 2003	BPI DSM-IV-TR (rec: M or m)	CGI-S <3	175	46	N
Calabrese et al. [22] 2003	BPI DSM-IV-TR (rec: D)	CGI-S <3	463	121	N
Calabrese et al. [23] 2005	BPI, BPII DSM-IV-TR Eu (his: cycl, rec: M/m/MX)	HRSD-24 <20, YMRS <12, GAS >51	60 (24 BPI, 36 BPII)	32 (19 BPII, 13 BPI)	Y Li, N DVP
Tohen et al. [25] 2005	BPI DSM-IV Eu (rec: M)	YMRS >20	431	214	Y
Findling et al. [24] 2005	Ped. BPI, BPII DSM-IV (rec: M/m)	CDRS-R, YMRS, C-GAS	60 (55 BPI, 5 BPII) (rap/mix/psy)	30 (28 BPI, 2 BPII)	Y

MRS = Mania Rating Scale; DSS = Depressive Syndrome Scale; GAS = Global Assessment Scale; YMRS = Young Mania Rating Scale; HDRS = Hamilton Rating Scale for Depression; CDRS-R = Children's Depressive Rating Scale – Revised; Eu = euthymia; rec = recently; his = history; cycl = rapid cyclicity; M = manic; m = hypomanic; MX = mixed; D = depressed; rap = rapid cycling; mix = mixed episode; psy = psychotic symptoms.

to lack evidence-based data on the use of lithium that may help the clinician in everyday practice. Besides considering diagnosis, last mood episodes (recently manic/hypomanic/mixed or depressed) before recruitment are also taken into consideration in order to select patients. Almost all studies include recently manic/hypomanic patients; recently depressed patients are considered in the study by Calabrese et al. only [22]. Last episode polarity does not seem to affect lithium efficacy, as shown by the two studies on LMT, in which lithium was consecutively tested in recently manic and recently depressed patients. The pediatric population suffers from a substantial lack of data on lithium efficacy as maintenance therapy for BD, and, given the supposedly long-term treatment in this specific subgroup of patients, further studies are urged. Rating scales have been used in order to quantify severity of symptoms and to define remission before patients entered the double-blind phase: LMT studies just quantify a global clinical impression. Generally mildly ill patients with good functioning were enrolled in the double-blind phases, as open-label phases are supposed to stabilize and select patients responsive and tolerant to treatment. Exclusion criteria such as axis I or II comorbidities are known to limit patient heterogeneity in every RCT: this may limit generalizability and, indeed, the effectiveness of lithium maintenance estimated by these trials may be superior to what is seen in everyday clinical practice [19]. It is widely accepted that lithium possesses a specific antisuicidal effect [39]. This aspect was not evaluated by RCTs: on the contrary, patients with a high risk for suicide were excluded.

The primary efficacy endpoints (table 3) used in studies were: *time to intervention* (addition to electroconvulsive therapy or pharmacotherapy, including antidepressants, antipsychotics, anticonvulsants/mood stabilizers) or *time to any mood episode* (manic, hypomanic, mixed or depressive). The threshold for intervention (need for pharmacotherapy or electroconvulsive therapy) is based on the investigators' overall clinical judgment of relapse and, in 2 studies [23, 25], on total scores on specific rating scales. Secondary efficacy endpoints include more specific BD characteristics, such as time to intervention for a manic, hypomanic or mixed episode, time to intervention for a depressive episode, mean change from baseline on rating scales' total scores. Considering the primary aim, lithium simply showed its higher efficacy compared to placebo, confirming clinical practice. The authors failed to highlight any statistical differences in efficacy compared to other sponsored drugs (DVP, LMT, OLZ). When specific polarity of episodes and/or mean change from baseline in rating scales' total scores (syndromic and symptomatic secondary aims) are considered, lithium shows antimanic properties profiling as a prototype of from-above mood stabilizers and does not reach statistically significant results in depressive episode prevention; it has not exhibited evidence of worsening of any phase of the illness neither in time to intervention nor, for secondary efficacy measures, on mean change in scores or proportion of patients with clinically significant worsening [22].

Table 3. Efficacy endpoints, primary and secondary outcomes, safety outcomes

References	Efficacy endpoints			Outcomes				Safety outcomes	
	primary outcome definition		secondary outcome definition	primary results		secondary results		Li vs. PBO (p value)	Li vs. drug (p value)
	clinical	relapse/ recurrence		Li vs. PBO (p value)	Li vs. drug (p value)	Li vs. PBO (p value)	Li vs. drug (p value)		
Bowden et al. [20] 2000	Tti any mood episode	clinical	(1) TtM (MRS >16 + hosp) (2) TtD (AD intervention) (3) mean change from baseline at MRS, DSS, GAS	–	0.05	–	MRS: 0.05 DSS: 0.05	tremor: 0.001 nausea: 0.05 diarrhea: 0.02	polyuria, polydipsia: 0.001 tachycardia, akathisia: 0.04 dry eyes: 0.03 MINOR: sedation: 0.02 infections: 0.009 tinnitus: 0.01
Bowden et al. [21] 2003	Tti any mood episode	need for pharmacotherapy or ECT, clinically assessed	(1) time to any disc (2) TtI for M/m/MX (3) TtI for D (4) mean change from baseline at MRS, HDRS, CGI-S, CGI-I, GAS	0.003	0.46	M/m: 0.006 D: 0.17 MRS: 0.001	TtI for D: 0.36 MRS: 0.03	diarrhea: 0.009 nausea: 0.009	diarrhea: 0.002 nausea: 0.02
Calabrese et al. [22] 2003	Tti any mood episode	need for pharmacotherapy or ECT, clinically assessed	(1) TtI for M/m/MX (2) TtI for D (3) mean change from baseline at MRS, HDRS, CGI-S, CGI-I, GAS	any: 0.029 D: 0.21	any: 0.915 D: 0.43 M/m: 0.12	M/m: 0.026 HDRS: <0.05 GAS: <0.05 CGI-S: <0.05	>0.05	somnolence: <0.05 tremor: <0.05	diarrhea: <0.05 tremor: <0.05
Calabrese et al. [23] 2005	Tti any mood episode	YMRS >20 (8 weeks) HDRS-24 >20 (8 weeks)	(1) time to full study disc (2) time to D or M/m/MX (3) mean change from baseline at YMRS, HDRS, GAS	–	any/M/m/D/MX: >0.05	–	any/M/m/D: >0.05	–	tremor: <0.02 polyuria/polydipsia: 0.001
Tohen et al. [25] 2005	incidence and time to any symp mood episode	YMRS and/ or HDRS >15	(1) symp recurrence or hosp for a mood episode (2) syndr (DSM-IV) recurrence after full remission	–	– any: symp/hosp: 0.02 syndr: 0.05 – M: YMRS: 0.02 RS/ hosp: 0.001 syndr: 0.001 – MX symp: 0.005 synt/ hosp: 0.02	–	D: 0.15 M: 0.02 MX: 0.005	–	<depression NAS: 0.01 >insomnia: <0.001 <hypersomnia: 0.03 >nausea: 0.2 <cholest: <0.001
Findling et al. [24] 2005	(1) TtI (2) disc for any reason	no definition given	mean change from baseline at CDRS-R, YMRS, CGI-S, CGI-I, CGAS	–	time to relapse: 0.05 disc for any reason: 0.72	–	N	–	>emesis: 0.05 >enuresis: 0.02 >polydipsia: 0.05

TtI = Time to intervention; M = manic; m = hypomanic; MX = mixed; D = depressed; disc = discontinuation; MRS = Mania Rating Scale; DSS = Depressive Syndrome Scale; GAS = Global Assessment Scale; HDRS = Hamilton Rating Scale for Depression; YMRS = Young Mania Rating Scale; CGAS = Children's Global Assessment Scale; CDRS-R = Children's Depressive Rating Scale – Revised; CGI-S = Clinical Global Inventory-Severity; CGI-I = Clinical Global Inventory-Improvement; ECT = electroconvulsive therapy; symp = symptomatic; syndr = syndromic; hosp = hospitalizations; cholest = cholesterol levels.

Safety Findings

Safety outcomes seem to record a significantly higher frequency of nausea, diarrhea, tremor and somnolence in the lithium group compared with placebo; compared to DVP, lithium shows less sedation, infections and tinnitus, but increases in polyuria and polydipsia, tachycardia, akathisia, and xerophthalmia. In pediatric patients, emesis, enuresis and polydipsia are more frequently recorded. When compared with LMT, lithium causes more diarrhea and tremor but less headache. Insomnia and nausea were more frequent in the lithium group compared to OLZ, while this drug showed more weight gain, increase in cholesterol levels and hypersomnia. Quite surprisingly, particularly feared adverse events in clinical practice, such as hypothyroidism or renal dysfunctions, are not significantly reported side effects during follow-ups.

Impact of Recent Data

At the time of finishing this report, a highly relevant study was presented for the first time during a meeting on January 25. Since it did not fit into the study period for our investigation, and since it has not undergone peer review as yet, we were not able to include it in this systematic review, but it is worth mentioning because it clearly supports most of our statements. This was a long-term, double-blind, placebo-controlled trial comparing quetiapine monotherapy and lithium monotherapy with placebo for the prevention of recurrences in patients who had previously been stabilized on quetiapine monotherapy [40]. The trial was enriched for quetiapine responders only and showed good results for quetiapine, but lithium also differed from placebo in preventing mania and depression. An important feature of this trial is that it is possibly the best quality trial proving truly *prophylactic* effects of lithium. While the enriched design speaks about relapse or recurrence prevention in responders to quetiapine, the fact that the trial was not enriched for lithium, as in the LMT maintenance studies, provides answers to the question of whether it may be worth using lithium as a prophylactic agent in patients who are already stable, even if they achieved stability by means of a different treatment (in this case, quetiapine). Importantly, in this trial, as opposed to the LMT studies, lithium was better than placebo not only in preventing mania, but also depression.

Conclusion

Data from randomized, double-blind, controlled studies are clearly supportive of the efficacy of lithium as a prophylactic agent against mania, and not as much against bipolar depression. Given the methodological shortcomings of most modern studies, in which lithium was generally included as an internal control for assay sensitivity but not assessed on a fair, head-to-head paradigm, it would be worthwhile to conduct further, well-designed trials to ascertain its current role as a prophylactic agent in manic-depressive illness. Its antisuicidal profile may be unique and deserves further investigation. Unanswered questions include its role in bipolar II disorder, in the elderly, and in combination with certain agents. This systematic review and the recent results of a monotherapy, placebo-controlled study indicate that lithium still has a very relevant role in preventing relapses and recurrences in BD, and that the decline of its prescription in several countries may not be a result due to evidence-based medicine, but to other factors involving pharmaceutical marketing and litigation.

References

1 Jefferson JW, Greist JH: Lithium; in Sadock BJ, Sadock VA (eds): Comprehensive Textbook of Psychiatry, ed 8. Philadelphia, Lippincott Williams & Wilkins, 2005, vol 2, pp 2839–2851.

2 Bech P: The full story of lithium. A tribute to Mogens Schou (1918–2005). Psychother Psychosom 2006;75:265–269.

3 Cade J: The story of lithium; in Ayd F, Blackwell B (eds): Discoveries in Biological Psychiatry. Philadelphia, Lippincott, 1970.

4 Schou M, Juel-Nielsen N, Stromgren E, Voldby H: The treatment of manic psychoses by the administration of lithium salts. J Neurol Neurosurg Psychiatry 1954;17:250–260.

5 Jope RS: Antibipolar therapy: mechanism of action of lithium. Mol Psychiatry 1999;4:117–128.

6 Manji HK, Potter WZ, Lenox RH: Signal transduction pathways. Arch Gen Psychiatry 1995;52:531–543.

7 Vieta E, Nolen WA, Grunze H, Licht RW, Goodwin G: A European perspective on the Canadian guidelines for bipolar disorder. Bipolar Disord 2005;7(suppl 3):73–76.

8 Wu RH, O'Donnell T, Ulrich M, Asghar SJ, Hanstock CC, Silverstone PH: Brain choline concentrations may not be altered in euthymic bipolar disorder patients chronically treated with either lithium or sodium valproate. Ann Gen Hosp Psychiatry 2004;3:13.

9 Serretti A: Lithium long-term treatment in mood disorders: clinical and genetic predictors. Pharmacogenomics 2002;3:117–129.

10 Ikonomov OC, Manji HK: Molecular mechanisms underlying mood stabilization in manic-depressive illness: the phenotype challenge. Am J Psychiatry 1999;156:1506–1514.
11 El-Mallakh S: Litihum: Actions and Mechanisms. Washington, American Psychiatric Press, 1996.
12 Manji HK, Bowden CL, Belmaker RH: Bipolar Medication. Mechanisms of Action. Washington, American Psychiatry Press, 1999.
13 Goodwin GM: Evidence-based guidelines for treating bipolar disorder: recommendations from the British Association for Psychopharmacology. J Psychopharmacol 2003;17:149–173; discussion 147.
14 Grunze H, Kasper S, Goodwin G, Bowden C, Baldwin D, Licht R, Vieta E, Möller HJ, World Federation of Societies of Biological Psychiatry Task Force on Treatment Guidelines for Bipolar Disorders: World Federation of Societies of Biological Psychiatry (WFSBP) guidelines for biological treatment of bipolar disorders. 1. Treatment of bipolar depression. World J Biol Psychiatry 2002;3:115–124.
15 Yatham LN, Kennedy SH, O'Donovan C, Parikh SV, MacQueen G, McIntyre RS, Sharma V, Beaulieu S, Guidelines Group, CANMAT: Canadian Network for Mood and Anxiety Treatments (CANMAT) guidelines for the management of patients with bipolar disorder: update 2007. Bipolar Disord 2006;8:721–739.
16 Hirschfeld RM, Montgomery SA, Aguglia E, Amore M, Delgado PL, Gastpar M, Hawley C, Kasper S, Linden M, Massana J, Mendlewicz J, Möller HJ, Nemeroff CB, Saiz J, Such P, Torta R, Versiani M: Partial response and nonresponse to antidepressant therapy: current approaches and treatment options J Clin Psychiatry 2002;63:826–837.
17 Tondo L, Hennen J, Baldessarini RJ: Lower suicide risk with long-term lithium treatment in major affective illness: a metaanalysis. Acta Psychiatr Scand 2001;104:163–172.
18 Moncrieff J: Lithium revisited. A re-examination of the placebo-controlled trials of lithium prophylaxis in manic-depressive disorder. Br J Psychiatry 1995;167:569–574.
19 Geddes JR, Burgess S, Hawton K, Jamison K, Goodwin GM: Long-term lithium therapy for bipolar disorder: systematic review and meta-analysis of randomized controlled trials. Am J Psychiatry 2004;161:217–222.
20 Bowden CL, Calabrese JR, McElroy SL, Gyulai L, Wassef A, Petty F, Pope HG Jr, Chou JC, Keck PE Jr, Rhodes LJ, Swann AC, Hirschfeld RM, Wozniak PJ: A randomized, placebo-controlled 12-month trial of divalproex and lithium in treatment of outpatients with bipolar I disorder. Divalproex Maintenance Study Group. Arch Gen Psychiatry 2000;57:481–489.
21 Bowden CL, Calabrese JR, Sachs G, Yatham LN, Asghar SA, Hompland M, Montgomery P, Earl N, Smoot TM, DeVeaugh-Geiss J, Lamictal 606 Study Group: A placebo-controlled 18-month trial of lamotrigine and lithium maintenance treatment in recently manic or hypomanic patients with bipolar I disorder. Arch Gen Psychiatry 2003;60:392–400. Erratum in: Arch Gen Psychiatry 2004;61:680.
22 Calabrese JR, Bowden CL, Sachs G, Yatham LN, Behnke K, Mehtonen OP, Montgomery P, Ascher J, Paska W, Earl N, DeVeaugh-Geiss J, Lamictal 605 Study Group: A placebo-controlled 18-month trial of lamotrigine and lithium maintenance treatment in recently depressed patients with bipolar I disorder. Clin Psychiatry 2003;64:1013–1024.
23 Calabrese JR, Shelton MD, Rapport DJ, Youngstrom EA, Jackson K, Bilali S, Ganocy SJ, Findling RL: A 20-month, double-blind, maintenance trial of lithium versus divalproex in rapid-cycling bipolar disorder. Am J Psychiatry 2005;162:2152–2161.
24 Findling RL, McNamara NK, Youngstrom EA, Stansbrey R, Gracious BL, Reed MD, Calabrese JR: Double-blind 18-month trial of lithium versus divalproex maintenance treatment in pediatric bipolar disorder. J Am Acad Child Adolesc Psychiatry 2005;44:409–417.
25 Tohen M, Greil W, Calabrese JR, Sachs GS, Yatham LN, Oerlinghausen BM, Koukopoulos A, Cassano GB, Grunze H, Licht RW, Dell'Osso L, Evans AR, Risser R, Baker RW, Crane H, Dossenbach MR, Bowden CL: Olanzapine versus lithium in the maintenance treatment of bipolar disorder: a 12-month, randomized, double-blind, controlled clinical trial. Am J Psychiatry 2005;162:1281–1290.
26 Hadjipavlou G, Mok H, Yatham LN: Pharmacotherapy of bipolar II disorder: a critical review of current evidence. Bipolar Disord 2004;6:14–25.
27 Rybakowski J: Long-term pharmacological treatment of bipolar disorders. Neuro Endocrinol Lett 2007;28(suppl 1):71–93.
28 Fountoulakis KN, Grunze H, Panagiotidis P, Kaprinis G: Treatment of bipolar depression: an update. J Affect Disord 2008;109:21–34.
29 Fountoulakis KN, Vieta E: Treatment of bipolar disorder: a systematic review of available data and clinical perspectives. Int J Neuropsychopharmacol 2008;11:999–1029.
30 Beynon S, Soares-Weiser K, Woolacott N, Duffy S, Geddes JR: Pharmacological interventions for the prevention of relapse in bipolar disorder: a systematic review of controlled trials. J Psychopharmacol 2009;23:574–591.
31 Dunner DL, Stallone F, Fieve RR: Lithium carbonate and affective disorders. 5. A double-blind study of prophylaxis of depression in bipolar illness. Arch Gen Psychiatry 1976;33:117–120.
32 Prien RF: NIMH report. Five-center study clarifies use of lithium, imipramine for recurrent affective disorders. Hosp Community Psychiatry 1984;35:1097–1098.
33 Prien RF, Caffey EM Jr, Klett CJ: Prophylactic efficacy of lithium carbonate in manic-depressive illness. Report of the Veterans Administration and National Institute of Mental Health collaborative study group. Arch Gen Psychiatry 1973;28:337–341.
34 Prien RF, Klett CJ, Caffey EM Jr: Lithium carbonate and imipramine in prevention of affective episodes. A comparison in recurrent affective illness. Arch Gen Psychiatry 1973;29:420–425.
35 Prien RF, Kupfer DJ, Mansky PA, Small JG, Tuason VB, Voss CB, Johnson WE: Drug therapy in the prevention of recurrences in unipolar and bipolar affective disorders. Report of the NIMH Collaborative Study Group comparing lithium carbonate, imipramine, and a lithium carbonate-imipramine combination. Arch Gen Psychiatry 1984;41:1096–1104.
36 Deshauer D, Fergusson D, Duffy A, Albuquerque J, Grof P: Re-evaluation of randomized control trials of lithium monotherapy: a cohort effect. Bipolar Disord 2005;7:382–387.
37 Vieta E, Carné X: The use of placebo in clinical trials on bipolar disorder: a new approach for an old debate. Psychother Psychosom 2005;74:10–16.
38 Uhlenhuth EH, Matuzas W, Warner TD, Thompson PM: Growing placebo response rate: the problem in recent therapeutic trials? Psychopharmacol Bull 1997;33:31–39.
39 Cipriani A, Pretty H, Hawton K, Geddes JR: Lithium in the prevention of suicidal behavior and all-cause mortality in patients with mood disorders: a systematic review of randomized trials. Am J Psychiatry 2005;162:1805–1819.
40 Nolen WA, Weisler RH, Neijber A, Hellqvist A, Paulsson B: Quetiapine or lithium versus placebo for maintenance treatment of bipolar I disorder after stabilization with quetiapine. 17th Eur Congr Psychiatry, Lisbon, January 24–28, 2009.

Neuropsychobiology

Neuropsychobiology 2010;62:36–42
DOI: 10.1159/000314308

Published online: May 7, 2010

Lithium's Emerging Role in the Treatment of Refractory Major Depressive Episodes: Augmentation of Antidepressants

Michael Bauer[a, b] Mazda Adli[b, c] Tom Bschor[a, b, d] Maximilian Pilhatsch[a]
Andrea Pfennig[a, b] Johanna Sasse[a, b] Rita Schmid[e] Ute Lewitzka[a, b]

[a]Department of Psychiatry and Psychotherapy, University Hospital Carl Gustav Carus, Technische Universität Dresden, Dresden, [b]The International Group for the Study of Lithium-Treated Patients, IGSLI, [c]Department of Psychiatry and Psychotherapy, Charité – Universitätsmedizin Berlin, Charité Campus Mitte, and [d]Department of Psychiatry, Schlosspark-Clinic, Berlin, and [e]Department of Psychiatry and Psychotherapy, University Medical Center Regensburg, Regensburg, Germany

Key Words
Lithium · Augmentation treatment strategies · Antidepressants · Treatment-resistant depression

Abstract
Background: The late onset of therapeutic response and a relatively large proportion of nonresponders to antidepressants remain major concerns in clinical practice. Therefore, there is a critical need for effective medication strategies that augment treatment with antidepressants. ***Methods:*** To review the available evidence on the use of lithium as an augmentation strategy to treat depressive episodes. ***Results:*** More than 30 open-label studies and 10 placebo-controlled double-blind trials have demonstrated substantial efficacy of lithium augmentation in the acute treatment of depressive episodes. Most of these studies were performed in unipolar depression and included all major classes of antidepressants, however mostly tricyclics. A meta-analysis including 10 randomized placebo-controlled trials has provided evidence that lithium augmentation has a statistically significant effect on the response rate compared to placebo with an odds ratio of 3.11, which corresponds to a number-needed-to-treat of 5. The meta-analysis revealed a mean response rate of 41.2% in the lithium group and 14.4% in the placebo group. One placebo-controlled trial in the continuation treatment phase showed that responders to acute-phase lithium augmentation should be maintained on the lithium-antidepressant combination for at least 12 months to prevent early relapses. Preliminary studies to assess genetic influences on response probability to lithium augmentation have suggested a predictive role of the –50T/C single nucleotide polymorphism of the GSK3β gene. ***Conclusion:*** Augmentation of antidepressants with lithium is currently the best-evidenced augmentation therapy in the treatment of depressed patients who do not respond to antidepressants.

History of Lithium in Depression

Since the discovery of lithium in modern psychiatry in 1949, lithium has been mainly considered as a prophylactic treatment for bipolar disorders and an acute treatment of mania. Less attention has been given to lithium

KARGER

Fax +41 61 306 12 34
E-Mail karger@karger.ch
www.karger.com

0302–282X/10/0621–0036$26.00/0

Accessible online at:
www.karger.com/nps

Michael Bauer, MD, PhD
Department of Psychiatry and Psychotherapy, University Hospital Carl Gustav Carus
Technische Universität Dresden, Fetscherstrasse 74, DE–01307 Dresden (Germany)
Tel. +49 351 458 2772, Fax +49 351 458 4324
E-Mail michael.bauer@uniklinikum-dresden.de

and its potential value to treat and prevent acute depressive episodes in unipolar depression [1, 2]. As early as in the 1960s and 70s, the antidepressant effect of lithium was investigated by numerous studies. Although there is quite some evidence from these studies for an antidepressive efficacy, lithium has never been used broadly as an antidepressant medication in unipolar and bipolar depressed patients in clinical practice. The maintenance properties of lithium were investigated since the late 1950s in both bipolar and unipolar affective illness. Many of these studies which focused on bipolar disease also had substantial data on the prophylactic effects of lithium in preventing the recurrence of unipolar depression. A recent comprehensive review of these studies shows that lithium prevents the recurrence of unipolar depression [1]. However, despite these findings the evidence failed to convince practitioners to use lithium to treat acute and prophylactic depressive disorders. This might be due to its relatively narrow therapeutic range of application and the necessity with lithium treatment to monitor lithium serum level regularly. The focus of this article is on another effective application of lithium in the treatment of depressive disorders that emerged since the 1980s: lithium augmentation of antidepressants.

Augmentation Therapy in Depression

Although there are many drugs available for the treatment of major depression, the overall treatment outcome of depressed patients is usually far from optimal. Regardless of the initial choice of antidepressant, about 30–50% of patients with a major depressive episode will not respond sufficiently to adequately performed first-line treatment [3]. Numerous treatment strategies have been described for use in antidepressant nonresponder and treatment-resistant depression [4, 5]. Augmentation treatment strategies involve adding a second drug other than an antidepressant to the treatment regimen when no response or only partial response has been achieved, with the goal of enhancing treatment. Augmentation agents include lithium, thyroid hormones, atypical neuroleptics, anticonvulsants, stimulants and buspirone. One advantage of augmentation is that it eliminates the period of transition between 1 antidepressant to another and builds on the partial response. Consequently, when they work, augmentation strategies can have a rapid effect. Secondly, augmentation is of benefit for patients who have had some response and may be reluctant to risk losing that improvement [3].

Of these strategies, lithium augmentation therapy is among the best-studied ones [6]. Lithium salts have been used to augment the efficacy of antidepressant medications for more than 25 years. The first study to test this hypothesis in patients with major depression was performed by de Montigny et al. [7]. They reported a dramatic response within 48 h to the addition of lithium in 8 patients who had not responded to at least 3 weeks of treatment with tricyclic antidepressants. The efficacy of the combination and rapidity of response has led many clinical research groups to pursue study of this treatment intervention. Subsequent randomized controlled trials have confirmed de Montigny's initial findings from 1981 with more than 30 open-label and comparator studies including more than 500 depressed patients being published [8]. In these studies, the duration of antidepressant before treatment ranged between 3 and 7 weeks with a mean of 4.5 weeks; the subsequent lithium augmentation therapy lasted between 2 days and 14 weeks with a mean duration of about 30 days. The antidepressants used in the trials included agents from different groups; among them were selective serotonin reuptake inhibitors (SSRIs), tri- and tetracyclic antidepressants, and monoamine oxidase inhibitors. The dosages of the antidepressants used were not reported in all trials. The dosages of lithium carbonate ranged between 300 and 1,500 mg/day. The response rates ranged widely between 100 and 23.5% with a median of 56%; 10 of 17 open-label studies found response rates to lithium augmentation of 50% or more.

A recent meta-analysis addressing the question pooled 10 randomized, double-blind, placebo-controlled trials and included 269 mostly unipolar depressed patients [9]. Lithium dosage, duration of treatment and other characteristics of the studies are detailed in table 1. Lithium had a significant positive effect versus placebo with an odds ratio of 3.11, which corresponds to a number-needed-to-treat of 5. Table 1 shows the odds ratio for subjects responding to the treatment in each study by year of publication.

However, 5 of the studies pooled did not show a significant difference. Reasons for the negative findings may be low power [10–12], use of insufficient lithium doses [13], too short a duration of treatment [10, 12], and concerns about the efficacy of lithium augmentation with noradrenergic antidepressants [14, 15]. Previous studies had demonstrated that only doses of lithium carbonate higher than 600 mg/day and a duration of 7 days were useful in augmenting therapies [16].

Table 1. Results of randomized, placebo-controlled lithium augmentation studies in depression

Study	Subjects	Antidepressant	Lithium dosage (serum level) and duration	Fixed effects: odds ratio[a] (95% CI)	Response criteria; response rates
Heninger et al. [47], 1983	14 UP, 1 BP 12 F, 3 M mean age 50 years	various TCAs and tetracyclics	lithium carbonate 900–1,200 mg/day (0.5–1.1 mEq/l) 12 days	23.57 (1.00–556.08)	decrease of 2 or more points on SCRS; lithium: 62.5% placebo: 0%
Kantor et al. [10], 1986	7 UP sex nr mean age nr	various TCAs	lithium carbonate 900 mg/day 48 h	3.00 (0.09–102.05)	≥40% decrease in HAM-D score; lithium: 25% placebo: 0%
Zusky et al. [11], 1988	16 UP 13 F, 3 M mean age 45 years	various TCAs and MAOIs	lithium carbonate 300 mg/day first week, 900 mg/day second week 14 days	1.80 (0.21–15.41)	final HAM-D score ≤7; lithium: 38% placebo: 25%
Schöpf et al. [30], 1989	18 UP, 9 BP 19 F, 8 M mean age 54 years	various antidepressants	lithium carbonate 600–800 mg/day (0.6–0.8 mEq/l) 7 days	27.00 (1.35–541.57)	≥50% decrease in HAM-D; lithium: 50% placebo: 0%
Browne et al. [12], 1990	14 UP, 3 BP 10 F, 7 M mean age 42 years	various TCAs and tetracyclics	lithium carbonate 900 mg/day 48 h	3.00 (0.35–25.87)	≥50% decrease in HAM-D; lithium: 43% placebo: 20%
Stein and Bernadt [13], 1993	34 UP 27 F, 7 M mean age 47 years	various TCAs	lithium carbonate 250 mg/day or 750 mg/day 21 days	0.50 (0.08–3.19)	≥50% decrease in HAM-D; lithium (250 mg): 18% lithium (750 mg): 44% placebo: 22%
Joffe et al. [21], 1993	33 UP 18 F, 15 M mean age 37 years	various TCAs	lithium carbonate 900 mg/day (>0.55 mEq/l) 14 days	4.88 (1.01–23.57)	≥50% decrease in HAM-D; lithium: 52% placebo: 18.7%
Katona et al. [48], 1995	61, polarity nr 35 F, 26 M mean age 40 years	SSRI and TCA	lithium 800 mg/day (0.6–1 mmol/l) 42 days	3.21 (1.09–9.48)	≥50% decrease in HAM-D; lithium: 53% placebo: 25%
Baumann et al. [49], 1996	23 UP, 1 BP 17 F, 7 M mean age 41 years	SSRI (citalopram)	lithium carbonate 800 mg/day (0.5–0.8 mmol/l) 7 days	9.00 (1.27–63.89)	≥50% decrease in HAM-D; lithium: 58% placebo: 14%
Nierenberg et al. [14], 2003	35 UP 16 F, 19 M mean age 38 years	TCA (nortriptyline)	lithium carbonate 900 mg/day	0.58 (0.08–4.01)	≥50% decrease in HAM-D; lithium: 12.5% placebo: 20%

UP = Unipolar; BP = bipolar; F = female; M = male; nr = not reported; MAOI = monoamine oxidase inhibitor; TCA = tricyclic antidepressant; HAM-D = Hamilton Depression Rating Scale; SCRS = Short Clinical Rating Scale; CI = confidence interval.
[a] All studies: odds ratio 3.11 (1.80–5.37); data from meta-analysis.

Since the first meta-analysis [16], only 1 placebo-controlled study has been published [14]. As noted on the original meta-analysis published in 1999 [16], a new negative study would have to include more than 2,500 patients per group to change the results of this pooling. However, it remains to be examined whether the response to lithium augmentation represents true augmentation resulting from synergistic effects or whether the response is simply owing to the antidepressant effect of lithium itself. Experimental studies supporting the former possibility will be reviewed below. From the clinical point of view, arguments for a true augmentation effect derive from a controlled clinical trial showing that the antidepressant effects of lithium addition were significantly higher in amitriptyline-pretreated depression patients, compared with placebo-pretreated patients, who showed no improvement after a 3-week treatment [17]. In summary, a randomized, double-blind study is warranted which investigates the effects of lithium alone and compares them with the effects of lithium in combination with an antidepressant.

Comparator Studies of Lithium Augmentation

Rarely have two augmentation strategies been compared with each other under similar conditions, thus the relative magnitude of their effects is largely unknown. Among the comparators were electroconvulsive therapy [18], monoamine oxidase inhibitors [19, 20], thyroid hormone [21], carbamazepine [22, 23], and high-dose SSRIs or tricyclic augmentation in SSRI users [24]. In these studies, there has been no dramatic difference between lithium augmentation and any other strategy, although the varied quality of some studies – for example problems with trial duration, use of subtherapeutic doses of lithium and low power – should preclude drawing definite conclusions.

Continuation Phase and Discontinuation Studies

One randomized controlled trial has examined the efficacy of lithium augmentation in the continuation treatment phase of unipolar major depressive disorder [25]. Twenty-nine patients with a refractory major depressive episode, single or recurrent, who had responded to acute lithium augmentation therapy during an open-label 6-week study were randomized after a 2- to 4-week stabilization period to a double-blind continuation treatment for another 4 months with either lithium (n = 14) or placebo (n = 15). The antidepressant was continued at the same dosage throughout the study. Seven of the 15 patients on placebo suffered from a relapse (5 depressive and 2 with a first manic episode) in the double-blind study phase, while no patients from the lithium group relapsed. Even more patients relapsed during the subsequent open-label 6-month phase after lithium was withdrawn in the group previously receiving lithium [26]. It was concluded that patients who respond to lithium augmentation should be maintained on lithium augmentation for a minimum of 12 months or even longer [26].

The effects of gradual discontinuation of lithium augmentation therapy were assessed in a randomized, placebo-controlled discontinuation study in 12 elderly patients who had responded to lithium augmentation during their most recent refractory unipolar depressive episode [27]. Patients were randomized to receive continued lithium augmentation or matching placebo; 2 of 6 patients in the lithium maintenance group had a recurrence of depression at 61 and 96 weeks, respectively, immediately after a stressful life event. Similarly, 2 of 6 patients had a recurrence in the placebo group at 7 and 92 weeks, respectively, without any apparent changes in life stresses. In a second, naturalistic discontinuation study in an elderly group of patients with major depressive disorder, about half of the patients relapsed following discontinuation of lithium augmentation [28]. In a recent review article [29], it was concluded that there is a significant risk of relapse in elderly patients whose lithium augmentation therapy for unipolar depression is discontinued.

Predictors of Response and Practical Use of Lithium Augmentation

A series of studies have investigated clinical and biological factors in lithium augmentation trials to allow outcome prediction [17, 30–33]. Age and gender have consistently been found *not* to be associated with response. Some studies have observed that bipolar patients respond better than unipolar patients [34], but this has not been confirmed by others, likely because most patients enrolled in the augmentation studies suffered from unipolar depression [35]. An analysis of 71 depressed patients refractory to treatment with a tricyclic antidepressant demonstrated that patients with a more severe depressive syndrome were more likely to respond to lithium augmentation [33].

During the past decade, we have administered lithium augmentation in a fast titrating regimen in adults without major adverse events in depressed in- and outpatients. In adults (age up to 55 years; in elderly patients slower), lithium carbonate is started on a daily dose of 450 mg (equivalent to approximately 12 mmol/day), and the dose is increased to 900 mg on the second day. This scheme leads to lithium serum levels of 0.5–0.7 mmol/l in most patients. The first dose adjustment can be performed after achieving a steady state, typically 5 days after the last change in dosing. Treatment of about 4 weeks allows assessment of the patient's response.

Mechanisms of Lithium Augmentation

The underlying mechanisms of action involved in the potentiatory effect of lithium is still unclear and several hypotheses have been suggested such as activity and modulation on serotonin (5-HT) neurotransmission and endocrine systems [36–38]. Initially, de Montigny and colleagues [17, 39] postulated that a pharmacodynamic action mediated via the serotonergic systems may account for the synergistic effect of lithium when added to a tricyclic antidepressant. From animal studies there is robust evidence that lithium augmentation increases 5-HT neurotransmission, possibly through a synergistic action of lithium and the antidepressant on brain 5-HT pathways. Using microdialysis techniques in animals, addition of lithium to chronic citalopram therapy further elevated basal levels of 5-HT in the rat ventral hippocampus [40]. Further evidence for a 'true' augmentation effect derived from animal studies showed that, in contrast to lithium alone, the addition of lithium to antidepressant treatment with the SSRI, citalopram potentiated presynaptic serotonergic function in rats [41]. Specifically, some authors suggested that the effect of lithium on 5-HT neurotransmission could be linked to a partial agonist activity on 5-HT1B autoreceptors or to a modulatory activity on these receptors [38]. In a recent series of experiments, the effects of lithium were investigated in a battery of standard behavioral tests (e.g. Porsolt's forced swim test, open-field test) that are considered animal models of depression. In these studies, behavior was robustly affected dose dependently by chronic lithium treatment [42, 43] suggesting lithium's genuine antidepressant properties. Remarkably, these lithium-sensitive behaviors were also observed in mice lacking one copy of the gene encoding GSK3β, a well-established direct target of lithium [42].

Pharmacogenetic approaches to assess genetic influences on response probability to lithium augmentation have also revealed some promising results. In a pilot study, the 5-HT transporter gene-linked polymorphic region (5-HTTLPR) allele variant's effect on lithium augmentation was analyzed in 50 antidepressant-nonresponsive patients [44]. Patients homozygous for the s allele of the 5-HTTLPR had a more favorable response compared with those heterozygous (hazard ratio = 6.9; $p = 0.005$) or homozygous for the l allele (hazard ratio = 4.5; $p = 0.003$). These findings support a differential effect of the 5-HTTLPR gene on treatment augmentation and the s/s genotype might predict an individual's risk of antidepressant nonresponsiveness and sensitivity to augmentative drugs such as lithium.

Another pharmacogenetic study investigated the association of the GSK3β –50T/C single nucleotide polymorphism (SNP) and response to lithium augmentation in 81 unipolar depressed patients [45]. Inhibition of GSK3β – an intracellular kinase, highly expressed in the brain – has been associated with neurotrophic and neuroprotective effects. Most interestingly, lithium is a direct and significant inhibitor of GSK3β. A –50T/C SNP localized within the promoter region of the GSK3β gene has previously been shown to be associated with response to lithium prophylaxis in bipolar disorder [46]. Similarly, carriers of the C allele of the –50T/C SNP showed a significantly better response to lithium augmentation (hazard ratio = 2.70, $p = 0.007$) with a mean remission rate of 56.2% after 4 weeks compared to 31% in patients with the TT genotype [45]. These results suggest a predictive role of the –50T/C SNP of the GSK3β gene for response to lithium augmentation treatment in depressed patients who do not respond sufficiently to monotherapy with an antidepressant. However, these promising data require confirmation in larger samples to be applicable in clinical routine care. Also, the functional relevance of the GSK3β –50T/C SNP needs to be further investigated.

Conclusion

Augmentation of antidepressants with lithium is the best-evidenced augmentation therapy in the treatment of depressed patients who do not respond to standard antidepressants. The evidence reviewed here supports the recommendation of lithium augmentation as a first-line therapy for nonresponding and refractory depressed patients [3, 4, 6]. In depressed patients who respond to lithium augmentation, effective lithium doses should be continued in combination with the antidepressant for at least 12 months after remission, in those patients with recurrent depression even longer [25]. Unfortunately, this effective and generally well-tolerated strategy is widely underutilized in clinical practice.

References

1 Davis JM: Lithium maintenance of unipolar depression; in Bauer M, Grof P, Müller-Oerlinghausen B (eds): Lithium in Neuropsychiatry – The Comprehensive Guide. Abingdon, Informa Healthcare UK Ltd, 2006, pp 99–108.

2 Bauer M, Crossley NA, Gerber S, Bschor T: The acute antidepressive effects of lithium: from monotherapy to augmentation therapy in major depression; in Bauer M, Grof P, Müller-Oerlinghausen B (eds): Lithium in Neuropsychiatry – The Comprehensive Guide. Abingdon, Informa Healthcare UK Ltd, 2006, pp 109–127.

3 Bauer M, Bschor T, Pfennig A, Whybrow PC, Angst J, Versiani M, Möller HJ: World Federation of Societies of Biological Psychiatry (WFSBP) guidelines for biological treatment of unipolar depressive disorders in primary care. World J Biol Psychiatry 2007;8:67–104.

4 Bauer M, Whybrow PC, Angst J, Versiani M, Möller HJ: World Federation of Societies of Biological Psychiatry (WFSBP) guidelines for biological treatment of unipolar depressive disorders. 1. Acute and continuation treatment of major depressive disorder. World J Biol Psychiatry 2002;3:5–43.
5 Thase ME: Augmentation strategies for depression: history and concepts. CNS Spectr 2007;12(suppl 22):3–5.
6 Carvalho AF, Machado JR, Cavalcante JL: Augmentation strategies for treatment-resistant depression. Curr Opin Psychiatry 2008; 22:7–12.
7 De Montigny C, Grunberg F, Mayer A, Deschenes JP: Lithium induces rapid relief of depression in tricyclic antidepressant drug non-responders. Br J Psychiatry 1981;138: 252–256.
8 Bauer M, Adli M, Baethge C, Berghöfer A, Sasse J, Heinz A, Bschor T: Lithium augmentation therapy in refractory depression: clinical evidence and neurobiological mechanisms. Can J Psychiatry 2003;48:440–446.
9 Crossley NA, Bauer M: Acceleration and augmentation of antidepressants with lithium for depressive disorders: two meta-analyses of randomized, placebo-controlled trials. J Clin Psychiatry 2007;68:935–940.
10 Kantor D, McNevin S, Leichner P, Harper D, Krenn M: The benefit of lithium carbonate adjunct in refractory depression – fact or fiction? Can J Psychiatry 1986;31:416–418.
11 Zusky PM, Biederman J, Rosenbaum JF, Manschreck TC, Gross CC, Weilberg JB, Gastfriend DR: Adjunct low-dose lithium carbonate in treatment-resistant depression: a placebo-controlled study. J Clin Psychopharmacol 1988;8:120–124.
12 Browne M, Lapierre YD, Hrdina PD, Horn E: Lithium as an adjunct in the treatment of major depression. Int Clin Psychopharmacol 1990;5:103–110.
13 Stein G, Bernadt M: Lithium augmentation therapy in tricyclic-resistant depression. A controlled trial using lithium in low and normal doses. Br J Psychiatry 1993;162:634–640.
14 Nierenberg AA, Papakostas GI, Petersen T, Montoya HD, Worthington JJ, Tedlow J, Alpert JE, Fava M: Lithium augmentation of nortriptyline for subjects resistant to multiple antidepressants. J Clin Psychopharmacol 2003;23:92–95.
15 Bschor T, Bauer M: Is successful lithium augmentation limited to serotonergic antidepressants? J Clin Psychopharmacol 2004;24: 240–241.
16 Bauer M, Döpfmer S: Lithium augmentation in treatment-resistant depression: meta-analysis of placebo-controlled studies. J Clin Psychopharmacol 1999;19:427–434.
17 de Montigny C, Cournoyer G, Morissette R, Langlois R, Caille G: Lithium carbonate neurobiologic actions of tricyclic antidepressant drugs and lithium ion on the serotonin system. Arch Gen Psychiatry 1983;40:1327–1334.
18 Dinan TG, Barry S: A comparison of electroconvulsive therapy with a combined lithium and tricyclic combination among depressed tricyclic nonresponders. Acta Psychiatr Scand 1989;80:97–100.
19 Hoencamp E, Haffmans PM, Dijken WA, Hoogduin CA, Nolen WA, van Dyck R: Brofaromine versus lithium addition to maprotiline. A double-blind study in maprotiline refractory depressed outpatients. J Affect Disord 1994;30:219–227.
20 Kok RM, Vink D, Heeren TJ, Nolen WA: Lithium augmentation compared with phenelzine in treatment-resistant depression in the elderly: an open, randomized, controlled trial. J Clin Psychiatry 2007;68:1177–1185.
21 Joffe RT, Singer W, Levitt AJ, MacDonald C: A placebo-controlled comparison of lithium and triiodothyronine augmentation of tricyclic antidepressants in unipolar refractory depression. Arch Gen Psychiatry 1993a;50: 387–393.
22 Rybakowski JK, Suwalska A, Chlopocka-Wozniak M: Potentiation of antidepressants with lithium or carbamazepine in treatment-resistant depression. Neuropsychobiology 1999;40:134–139.
23 Schüle C, Baghai TC, Eser D, Nothdurfter C, Rupprecht R: Lithium but not carbamazepine augments antidepressant efficacy of mirtazapine in unipolar depression: an open-label study. World J Biol Psychiatry 2008;25:1–10.
24 Fava M, Alpert J, Nierenberg A, Lagomasino I, Sonawalla S, Tedlow J, Worthington J, Baer L, Rosenbaum JF: Double-blind study of high-dose fluoxetine versus lithium or desipramine augmentation of fluoxetine in partial responders and nonresponders to fluoxetine. J Clin Psychopharmacol 2002;22: 379–387.
25 Bauer M, Bschor T, Kunz D, Berghöfer A, Ströhle A, Müller-Oerlinghausen B: Double-blind, placebo-controlled trial of the use of lithium to augment antidepressant medication in continuation treatment of unipolar major depression. Am J Psychiatry 2000;157: 1429–1435.
26 Bschor T, Berghöfer A, Ströhle A, Kunz D, Adli M, Müller-Oerlinghausen B, Bauer M: How long should the lithium augmentation strategy be maintained? A 1-year follow-up of a placebo-controlled study in unipolar refractory major depression. J Clin Psychopharmacol 2002;22:427–430.
27 Hardy BG, Shulman KI, Zucchero C: Gradual discontinuation of lithium augmentation in elderly patients with unipolar depression. J Clin Psychopharmacol 1997;17:22–26.
28 Fahy S, Lawlor BA: Discontinuation of lithium augmentation in an elderly cohort. Int J Geriatr Psychiatry 2001;16:1004–1009.
29 Ross J: Discontinuation of lithium augmentation in geriatric patients with unipolar depression: a systematic review. Can J Psychiatry 2008;53:117–120.
30 Schöpf J, Baumann P, Lemarchand T, Rey M: Treatment of endogenous depressions resistant to tricyclic antidepressants or related drugs by lithium addition. Results of a placebo-controlled double-blind study. Pharmacopsychiatry 1989;22:183–187.
31 Joffe RT, Levitt AJ, Bagby RM, MacDonald C, Singer W: Predictors of response to lithium and triiodothyronine augmentation of antidepressants in tricyclic non-responders. Br J Psychiatry 1993b;163:574–578.
32 Alvarez E, Perez-Sola V, Perez-Blanco J, Queralto JM, Torrubia R, Noguera R: Predicting outcome of lithium added to antidepressants in resistant depression. J Affect Disord 1997;42:179–186.
33 Bschor T, Canata B, Müller-Oerlinghausen B, Bauer M: Predictors of response to lithium augmentation in tricyclic antidepressant-resistant depression. J Affect Disord 2001;64: 261–265.
34 Rybakowski J, Matkowski K: Adding lithium to antidepressant therapy: factors related to therapeutic potentiation. Eur Neuropsychopharmacol 1992;2:161–165.
35 Price LH, Charney DS, Heninger GR: Variability of response to lithium augmentation in refractory depression. Am J Psychiatry 1986;143:1387–1392.
36 Chenu F, Bourin M: Potentiation of antidepressant-like activity with lithium: mechanism involved. Curr Drug Targets 2006;7: 159–163.
37 Bschor T, Adli M, Baethge C, Eichmann U, Ising M, Uhr M, Modell S, Künzel H, Müller-Oerlinghausen B, Bauer M: Lithium augmentation increases the ACTH and cortisol response in the combined DEX/CRH test in unipolar major depression. Neuropsychopharmacology 2002;27:470–478.
38 Bschor T, Baethge C, Adli M, Eichmann U, Ising M, Uhr M, Modell S, Künzel H, Müller-Oerlinghausen B, Bauer M: Association between response to lithium augmentation and the combined DEX/CRH test in major depressive disorder. J Psychiatr Res 2003;37: 135–143.
39 de Montigny C, Aghajanian GK: Tricyclic antidepressants: long-term treatment increases responsivity of rat forebrain neurons to serotonin. Science 1978;202:1303–1306.
40 Wegener G, Bandpey Z, Heiberg IL, Mork A, Rosenberg R: Increased extracellular serotonin level in rat hippocampus induced by chronic citalopram is augmented by subchronic lithium: neurochemical and behavioural studies in the rat. Psychopharmacology (Berl) 2003;166:188–194.

41 Okamoto Y, Motohasi N, Hayakawa H, Muraoka M, Yamawaki S: Addition of lithium to chronic antidepressant treatment potentiates presynaptic serotonergic function without changes in serotonergic receptors in the rat cerebral cortex. Neuropsychobiology 1996;33:17–20.
42 O'Brien WT, Harper AD, Jové F, Woodgett JR, Maretto S, Piccolo S, Klein PS: Glycogen synthase kinase-3β haploinsufficiency mimics the behavioral and molecular effects of lithium. J Neurosci 2004;24:6791–6798.
43 Bersudsky Y, Shaldubina A, Belmaker RH: Lithium's effect in forced-swim test is blood level dependent but not dependent on weight loss. Behav Pharmacol 2007;18:77–80.
44 Stamm TJ, Adli M, Kirchheiner J, Smolka MN, Kaiser R, Tremblay PB, Bauer M: Serotonin transporter gene and response to lithium augmentation in depression. Psychiatr Genet 2008;18:92–97.
45 Adli M, Hollinde DL, Stamm T, Wiethoff K, Tsahuridu M, Kirchheiner J, Heinz A, Bauer M: Response to lithium augmentation in depression is associated with the glycogen synthase kinase 3-beta -50T/C single nucleotide polymorphism. Biol Psychiatry 2007;62: 1295–1302.
46 Benedetti F, Serretti A, Pontiggia A, Bernasconi A, Lorenzi C, Colombo C, Smeraldi E: Long-term response to lithium salts in bipolar illness is influenced by the glycogen synthase kinase 3-beta -50 T/C SNP. Neurosci Lett 2005;376:51–55.
47 Heninger GR, Charney DS, Sternberg DE: Lithium carbonate augmentation of antidepressant treatment. An effective prescription for treatment-refractory depression. Arch Gen Psychiatry 1983;40:1335–1342.
48 Katona CL, Abou-Saleh MT, Harrison DA, Nairac BA, Edwards DR, Lock T, Burns RA, Robertson MM: Placebo-controlled trial of lithium augmentation of fluoxetine and lofepramine.Br J Psychiatry 1995;166:80–86.
49 Baumann P, Nil R, Souche A, Montaldi S, Baettig D, Lambert S, Uehlinger C, Kasas A, Amey M, Jonzier-Perey M: A double-blind, placebo-controlled study of citalopram with and without lithium in the treatment of therapy-resistant depressive patients: a clinical, pharmacokinetic, and pharmacogenetic investigation. J Clin Psychopharmacol 1996; 16:307–314.

Neuropsychobiology

Neuropsychobiology 2010;62:43–49
DOI: 10.1159/000314309

Published online: May 7, 2010

Lithium Reduces Pathological Aggression and Suicidality: A Mini-Review

Bruno Müller-Oerlinghausen[a] Ute Lewitzka[b]

[a]Drug Commission of the German Medical Association, Berlin, and [b]Department of Psychiatry and Psychotherapy, Technical University of Dresden, Dresden, Germany

Key Words

Lithium · Pathological aggression · Children · Prisoners · Suicide · Suicide attempt

Abstract

From a practical point of view, the well-proven antisuicidal and anti-aggressive effects of lithium are of utmost importance for a rational, safe and economical treatment of patients with affective disorders. Regular lithium long-term treatment reduces the otherwise 2- to 3-fold increased mortality of untreated patients with severe affective disorders down to the level of the general population. This is mainly due to the reduced suicide risk. Many international studies have confirmed this fascinating property of lithium which so far has not been demonstrated with comparable evidence for any other psychotropic compound. The antisuicidal effects of lithium might possibly be related to its anti-aggressive effects which have been shown in various species, populations and settings, such as animals, inhabitants of nursing homes for the elderly, mentally handicapped subjects, children and adolescents with hyperactive, hostile and aggressive behavior, and particularly in hyperaggressive inmates of correction units and prisons.

Introduction

Sixty years of lithium treatment in psychiatry yielded a still continuing stream of knowledge on the pharmacological and therapeutic effects as well as the effectiveness of this fascinating alkali ion. One of the lightest metallic elements had a heavy impact on the concepts and practice of modern psychiatry. Some scientific studies on lithium and clinical experience collected during recent decades partly confirmed early anecdotal findings, and partly remodeled and integrated them into updated, more precise and comprehensive concepts, thus making previous observations better understood [1, 2]. Other studies detected or redetected important aspects of lithium which either are still to be discussed on a hypothetical level such as its 'neuroprotective' effects, or which are supported by very good scientific evidence but still did not find their way into everyday clinical practice and existing guidelines. The antisuicidal and anti-aggressive properties of lithium reflect this latter situation. Although antidepressants have been in use for more than 50 years, it is disappointing that no valid data exist to prove a suicide- or mortality-reducing effect, with the consequence that usually drug regulatory agencies do not consider reducing the death rate as an essential criterion for approval of a new antidepressant or mood stabilizer. This is rather

KARGER

Fax +41 61 306 12 34
E-Mail karger@karger.ch
www.karger.com

0302–282X/10/0621–0043$26.00/0

Accessible online at:
www.karger.com/nps

Prof. Dr. med. Bruno Müller-Oerlinghausen
Bartningallee 11–13
DE–10557 Berlin (Germany)
Tel. +49 30 2336 2940, Fax +49 30 2336 2935
E-Mail bmoe@zedat.fu-berlin.de

astonishing since for lithium salts such data do exist. In what follows, we will sketch the key studies and the history of this important research area. More comprehensive reviews of the existing literature can be found elsewhere [3].

Does Regular Lithium Long-Term Treatment Diminish the Risk of Suicide in Patients with Affective Disorders?

Suicide Risk and Mortality of Patients with Affective Disorders

Patients with affective disorders exhibit a 2–3 times increased mortality when compared to the general population [4–7]. This excess mortality is caused primarily by the possibly 30- to 70-fold higher suicide-related mortality [8] – which is particularly high in patients with a history of suicide attempts [9, 10]. The meta-analysis by Guze and Robins [11] calculated the lifetime suicide risk as 15% for affective disorders, whereas 20 years later Goodwin and Jamison [12], based on more recent literature, reported an overall risk of 19%.

According to Harris and Barraclough [13], the suicide-related standardized mortality rate (SMR) [14] is 21.24 in major depression and 11.73 in bipolar disorder, with, however, large confidence intervals. Although some studies found a somewhat lower suicide rate in bipolar as compared to unipolar patients [15, 16], others found higher rates in bipolar patients [17–19], particularly in bipolar II patients.

The intriguing question whether long-term medication with lithium salts can improve the course of the manic-depressive disease or of affective disorders in terms of suicide prevention has been given little attention until the 80s.

Early Studies

Barraclough [20] was one of the first investigators postulating a potential association between lithium long-term medication and suicide prevention. Based on a detailed analysis of the charts of 100 suicide victims he concluded that about 20% of the suicides could have been prevented by adequate lithium medication. The first systematic retrospective study demonstrating a highly significant reduction of suicide attempts in a sample of 64 high-risk patients, 46 bipolar and 11 schizoaffective, during long-term lithium treatment was published in 1992 by the Berlin research group [21]. The authors emphasized that suicides and suicide attempts occurred nearly exclusively in a group of 13 patients who had taken lithium irregularly or had stopped the medication.

Felber's group in Dresden analyzing suicide attempts during accumulated periods on and off lithium had very similar findings: 90% of the suicide attempts occurred in the lithium-free period [22].

Several studies on the mortality of affective disorders during lithium long-term treatment by Coppen et al. [23] and by the International Group for the Study of Lithium-Treated Patients (IGSLI) [14, 24] demonstrated that the SMR of patients with affective disorders during adequate lithium medication is normalized down to the level of the general population[1]. Coppen [25] reviewed the studies existing in the mid-nineties on the suicide rates in patients on versus off lithium and concluded that adequate lithium medication reduces the suicide-related mortality by 82%.

The IGSLI Studies

In the IGSLI main study, well-documented data on the course of illness of 827 patients with affective disorders from lithium clinics in Austria, Canada, Denmark, and Germany, who had been treated with lithium for at least 6 months, were evaluated [14, 24]. Fifty-five percent of the patients were bipolar, 25% unipolar, 2% unipolar-manic, 16% schizoaffective, and 2% had other diagnoses. At onset of the lithium prophylaxis, patients were 41 years old on average. The mean duration of lithium treatment was 81 months (6–21 years), equaling 5,600 patient-years [14, 24].

The ratio of 44 observed and 38 expected cases of deaths is not statistically different from 1.0, which is the mortality of the general population. Thus, the expected 2- to 3-fold excess mortality in patients with affective disorders (see above) does no more exist in this lithium-treated patient sample.

Bipolar patients do not differ essentially from other diagnostic groups in this respect.

Although the specific suicide-related SMR is still higher than in the general population, it can be clearly shown that it is definitely lower in all diagnostic groups compared to what could be expected in untreated patient samples.

To weaken this finding, it might be speculated that patients willing to accept long-term treatment may exhibit a lower suicide risk, although this does not appear very likely when accounting for the fact that those patients would suffer from particularly severe courses of their disease. A further analysis of the IGSLI material could refute this argument: during the first year of lithium treatment, patients showed a 2-fold increased overall mortality and a 17-fold increased suicide-related mortality compared with the general population. During later lithium treatment, the SMR normalized thus indicating that patients who accept lithium prophylaxis actually exhibit a high suicide risk [26]. The Boston group also reported that after abrupt discontinuation of lithium medication the suicide risk rises particularly high [27].

Further Studies and Meta-Analyses

A meta-analysis of the data collected by the late nineties revealed that the risk of suicidal acts in bipolar patients off lithium was 7–8 times higher than in those on lithium [28]. Furthermore, the findings by Guzzetta et al. [29] support the reports by Coppen and the IGSLI that the antisuicidal effect also holds true for patients with recurrent depressions.

Linking regional lithium prescription rates with the Danish suicide register, Kessing et al. [30] added another impressive piece of evidence to the existing data on lithium's antisuicidal effect. There also exist convincing findings from controlled prospective studies. Cipriani et al. [31] reviewed 22 randomized controlled trials and found that patients on lithium were less likely to die from suicide, as compared to other compounds or placebo. Additional proof comes from the first controlled trial with lithium versus placebo as an adjunctive treatment in a large sample of high-risk patients. Suicidal events were defined as the endpoint [32]. Whereas the number of suicide attempts did not differ between treatment groups, a statistically significant superiority of lithium in terms of completed suicides was observed, whereas in a retrospective chart review of long-term treated bipolar patients from a single private practice nonfatal suicidal behavior was not different in patients on lithium and those on anticonvulsants [33].

Questions

There are several intriguing questions related to the mechanism of the antisuicidal effect of lithium and its potential specificity. First of all, is the antisuicidal effect coupled directly to the antidepressant effect of lithium? Several reports point to a mechanism separate from lithium's antidepressant action [34]. It has been speculated that the antisuicidal effect is associated with the well-proven serotonin-agonistic effect in biochemical terms, and with the well-established anti-aggressive effect of lithium in clinical terms [35]. Secondly, is the antisuicidal effect of lithium shared by other compounds such as antidepressants, neuroleptics or mood stabilizers? So far, in various trials using different methodologies, lithium has been shown clearly superior to other mood stabilizers in terms of suicide prevention [36–40].

Conclusion

'Does lithium save lives?' – the question raised in an editorial [41], can now be answered conclusively: yes, it does, if the treatment is applied and monitored in a rational, careful and individualized way. While there are rare negative findings in this respect [42], they can most likely be explained by incomplete patient follow-up and insufficient information on patients' compliance with lithium. This saving of lives considerably adds to the preference of lithium when the cost-effectiveness of treatment with various mood stabilizers is assessed [43].

Does Lithium Reduce Pathological Aggression?

The first report on the anti-aggressive effect of lithium was described by John F. Cade [44] who documented that lithium would be effective in cases of restless psychopathic mental defectives in a similar way as a prefrontal leukotomy.

Twenty years later, a number of Scandinavian case reports were published. Forssmann and Walinder [45] tried lithium for various indications. They reported 18 cases with aggression as the predominant clinical feature and pointed to amazingly good results with lithium.

Baastrup [46] characterized a female aggressive patient endangering herself and others. Lithium was the only drug to control this behavior although the patient was not manic-depressive.

Clinical trials on the effect of lithium on aggression included mainly four different groups of patients: (a) children with behavioral disturbances, (b) patients with mental handicaps, (c) adult psychiatric populations, and (d) persons with uncontrolled fit of rage. There is, however, some overlapping between these diagnostic groups.

Children with Behavioral Disturbances

The validity of studies conducted in children with aggressive behavior suffers from methodological problems. Most studies involved extremely heterogeneous patients. At least 4 double-blind, placebo-controlled studies were performed since 1995 [47–50].

Campell et al. [47] examined 61 children aged 5.2–12.9 years with severe aggressiveness, explosiveness and disruptiveness. The children had not responded to previous treatment. They were treated with lithium or haloperidol or placebo for 4 weeks. Both substances, lithium and haloperidol, reduced target symptoms such as hyperactivity, aggression and hostility, but side effects occurred significantly more often on haloperidol.

In the early 1990s, studies were often published which failed to replicate these results. The study of Carlson et al. [51] only showed 3 of 11 children with significant improvement as to self-control, aggression and irritability.

A second placebo-controlled double-blind trial was performed also by Campbell et al. [48]. Fifty children with severe aggressive and impulsive symptoms as part of a conduct disorder were treated with lithium or placebo for 6 weeks followed by 2 weeks on placebo. The findings were not as much in favor of lithium as those of the former study by Campell et al. [47].

Rifkin et al. [49] included 33 adolescent inpatients and administered lithium or placebo for 2 weeks. 8.3% of those who completed the study receiving placebo versus 21.4% of those receiving lithium were considered responders. The authors assumed the major limitation of this study and the possible reason for the nonsignificant results was the short period of treatment.

Another study conducted by Malone et al. [50] investigated different aggression subtypes related to treatment response. They included 28 aggressive conduct-disordered children (mean age 12.69 years) and treated them in a double-blind manner either with lithium or placebo. They found that treatment response with lithium correlated rather with the affective subtype than with the predatory subtype of aggression. A further study of the same authors [52] on the efficacy of lithium in the treatment of aggression achieved more favorable results for lithium. Lithium turned out to be statistically and clinically superior to placebo. They included 86 inpatients with conduct disorder and treated them in a double-blind manner with lithium or placebo for 4 weeks. Sixteen of 20 subjects in the lithium group were responders versus 6 of 20 in the placebo group. In more than 50% of the subjects in the lithium group adverse side effects like vomiting, nausea and increased urinary frequency were observed.

Patients with Mental Handicaps

Ten to thirty-five percent of patients with mental handicaps show self-injurious behavior like head-banging, biting, and scratching oneself. Explosive behavior complicates the daily life of these patients and their families and hospitalization, therefore, is often unavoidable. The treatment of aggressive behavior presents a challenge to every clinician.

Since the 1970s, lithium therapy was included in pharmacological attempts to reduce the most severe aggressive symptoms. Only studies in which the authors had clearly diagnosed the patients as mentally handicapped are listed in the following paragraph.

The first open trial on aggression was conducted by Dostal and Zvolsky [53]. They investigated 14 adolescent, mentally handicapped patients and showed a beneficial effect under lithium therapy. Most of the patients suffered from side effects like polydipsia and polyuria.

In an open study conducted by Micev and Lynch [54], these positive results could be confirmed. Six out of 8 patients experienced complete remission of self-harming behavior.

An equal reduction in aggression and self-injurious behavior was shown in a retrospective study by Luchins and Dojka [55] who assessed the effect of lithium and propanolol in subjects with mental retardation.

In 1989, Glenn et al. [56] published case reports of 10 brain-injured patients with severe aggressive behavior. Five of these patients showed a dramatic response after onset of lithium treatment.

A decrease in the frequency of aggressive outbursts was also demonstrated by Bellus et al. [57] in patients following brain injury. The authors concluded that the use of lithium may yield positive effects in the control of aggressive behavior, even in the long term.

Adult Psychiatric Population

One of the first studies in this field was conducted by van Putten and Sanders [58]. They treated 35 patients with chronic and incapacitating mental illness and reported a favorable change in aggressive behavior during lithium medication. None of their patients was suffering from manic-depressive illness.

Likewise remarkable results were found by Rifkin et al. [59]. This research group in 1972 conducted a 6-week double-blind crossover trial with random allocation comparing lithium to placebo in patients with emotionally unstable character disorder. In the view of Rifkin et al. [59], the persuasive improvement during the lithium treatment had to be explained either by hidden manic-depressive illness or by a closely related condition.

In 2001, Prado-Lima et al. [60] examined 8 mothers with child abuse. Lithium significantly reduced their abusive behavior towards their children at day 30 and day 60 of treatment compared with baseline. Especially the score for physical aggression towards other people showed a statistically significant reduction.

Adult patients with an attention-deficit/hyperactivity disorder were treated in a randomized double-blind crossover design by Dorrego et al. [61] with lithium or methylphenidate. They received 8 weeks of methylphenidate hydrochloride treatment and 8 weeks of lithium treatment by random assignment. Aggression was measured with the Overt Aggression Scale. The authors concluded that lithium might be beneficial in the treatment of adult attention-deficit/hyperactivity disorder and that lithium significantly alleviated a couple of behavioral problems including aggressive outbursts.

Persons with Uncontrolled Rage Outbursts

One interesting approach towards the effectiveness of lithium in the treatment of aggression is to study individuals with uncontrolled rage outbursts such as inmates of prisons or correctional institutions. In the 1970s, a number of 'prisoner studies' were conducted to investigate the effect of lithium on aggression.

Twelve inmates with personality disorders (aggressive, sociopathic and schizoid) from a maximum security prison in Connecticut were investigated by Sheard [62] who was the first to present study results comparing lithium with placebo. Lithium reduced self-rated aggressive affect while also being associated with a lower incidence of disciplinary sanctions.

Tupin et al. [63] treated 27 inmates of the California Medical Facility selected because of serious repeated violent behavior with lithium for 10 months. The anti-aggressive effect of lithium could be replicated. Participants of this study described an increased capacity to reflect on the consequences of actions and to control angry feelings while taking lithium.

The only clinical trial on lithium and aggression involving outpatients with uncontrolled fits of rage was conducted by Sheard [64]. Twelve inmates with aggressive behavior were treated with lithium for 4 months and after their release the patients were followed as outpatients during 1–2 months of lithium treatment. A decreased number of serious aggressive episodes could be shown.

The same authors conducted a 3-month, double-blind, placebo-controlled trial with 66 inmates from a medium-security prison [65]. Significantly fewer major infractions occurred in the lithium as compared to the placebo group. These results strengthen the evidence that lithium can have an inhibitory effect on impulsive aggressive behavior.

Potential Mechanism: The Role of Serotonin

The potential mechanisms of the anti-aggressive effects of lithium are unclear. Some authors [66] pointed to the frequently combined occurrence of impulse control disorder and bipolar disorder and suggested a relationship between both disorders. They discussed an underlying common pathophysiological pathway. Animal and clinical studies have suggested a critical role for central serotonin function in this context. Lithium exerts serotonin-enhancing effects [67] and this might provide an explanation for a possible working mechanism in the reduction of aggressive behavior.

Coccaro and colleagues [68–70] could show an inverse covariation between peripheral markers of central 5-HT functions and impulsive aggressive behavior such as a reduced number of 5-HT transporters on platelets correlated with aggressive behavior. However, the validity of studies allegedly corroborating these rather populist concepts has been questioned [71].

References

1 Grof P, Müller-Oerlinghausen B: A critical appraisal of lithium's efficacy and effectiveness: the last 60 years. Bipolar Disord 2009; 11(Suppl 2):10–19.
2 Grof P: Sixty years of lithium responders. Neuropsychobiology 2010;62:8–16.
3 Bauer M, Grof P, Müller-Oerlinghausen B (eds): Lithium in Neuropsychiatry. London, Informa, 2006, pp 33–42.
4 Lundquist G: Prognosis and cause of manic depressive psychosis. Acta Psychiatr Neurol Scand 1945;35:1–96.
5 Petterson U: Manic-depressive illness: a clinical, social and genetic study. Acta Psychiatr Scand Suppl 1977;269:55–60.
6 Tsuang MT, Woolson RF: Excess mortality in schizophrenia and affective disorders. Do suicides and accidental deaths solely account for this excess? Arch Gen Psychiatry 1978;35: 1181–1185.
7 Weeke A: Causes of death in manic depressives; in Schou M, Strömgren E (eds): Origin, Prevention and Treatment of Affective Disorders. London, Academic Press, 1979, pp 289–299.
8 Hagnell O, Lanke J, Rorsman B: Suicide rates in the Lundby study: mental illness as a risk factor for suicide. Neuropsychobiology 1981; 7:248–253.
9 Tuckman J, Youngman W: Identifying suicide with groups among attempted suicide. Public Health Rep 1963;78:763–766.
10 Motto I: Suicide attempts: a longitudinal view. Arch GenPsychiatry 1965;13:516–520.
11 Guze SB, Robins E: Suicide in primary affective disorders. Br J Psychiatry 1970;117:437–438.

12 Goodwin FK, Jamison KR: Manic-Depressive Illness. New York, Oxford University Press, 1990.
13 Harris EC, Barraclough B: Excess mortality of mental disorder. Br J Psychiatry 1998;173:11–53.
14 Ahrens B, Müller-Oerlinghausen B, Schou M, et al: Excess cardiovascular and suicide mortality of affective disorders may be reduced by lithium-prophylaxis. J Affect Disord 1995;33:67–75.
15 Newman S, Bland R: Suicide risk varies by subtype of affective disorder. Acta Psychiatr Scand 1991;83:420–426.
16 Ösby U, Brandt L, Correia N, et al: Excess mortality in bipolar and unipolar disorder in Sweden. Arch Gen Psychiatry 2001;58:844–850.
17 Lester D: Suicidal behavior in bipolar and unipolar affective disorders: a meta-analysis. J Affect Disord 1993;27:117–121.
18 Angst J, Sellaro R, Angst F: Long-term outcome and mortality of treated vs untreated bipolar and depressed patients: a preliminary report. Int J Psychiatry Clin Pract 1998;2:115–119.
19 Bottlender R, Jager M, Strauss A, et al: Suicidality in bipolar compared to unipolar depressed inpatients. Eur Arch Psychiatry Clin Neurosci 2000;250:257–261.
20 Barraclough B: Suicide prevention, recurrent affective disorder and lithium. Br J Psychiatry 1972;121:91–392.
21 Müller-Oerlinghausen B, Müser-Causemann B, Volk J: Suicides and parasuicides in a high-risk patient group on and off lithium long-term medication. J Affect Disord 1992;25:261–270.
22 Felber W, Kyber A: Suizide und Parasuizide während und ausserhalb einer Lithiumprophylaxe; in Müller-Oerlinghausen B, Berghöfer A (eds): Ziele und Ergebnisse der medikamentösen Prophylaxe affektiver Psychosen. Stuttgart, Thieme, 1994, pp 53–59.
23 Coppen A, Standish-Barry H, Bailey J, et al: Does lithium reduce the mortality of recurrent mood disorders? J Affect Disord 1991;23:1–7.
24 Müller-Oerlinghausen B, Ahrens B, Grof E, et al: The effect of long-term lithium treatment on the mortality of patients with manic-depressive and schizo-affective illness. Acta Psychiatr Scand 1992;86:218–222.
25 Coppen A: Depression as a lethal disease: prevention strategies. J Clin Psychiatry 1994;55(suppl 4):37–45.
26 Müller-Oerlinghausen B, Wolf T, Ahrens B, et al: Mortality during initial and during later lithium treatment. A collaborative study by the International Group for the Study of Lithium-treated Patients. Acta Psychiatr Scand 1994;90:295–297.
27 Baldessarini RJ, Tondo L, Hennen J: Effects of lithium treatment and its discontinuation on suicidal behaviour in bipolar manic-depressive disorders. J Clin Psychiatry 1999;60(suppl 2):77–84.
28 Baldessarini RJ, Tondo L, Hennen J: Effect of lithium treatment and its discontinuation on suicidal behavior in bipolar manic-depressive disorders. J Clin Psychiatry 1999;60(suppl 2):77–84.
29 Guzzetta F, Tondo L, Centorrino F, et al: Lithium treatment reduces suicide risk in recurrent major depressive disorder. J Clin Psychiatry 2007;68:380–383.
30 Kessing LV, Sondergard L, Kvist K, et al: Suicide risk in patients treated with lithium. Arch Gen Psychiatry 2005;62:860–866.
31 Cipriani A, Pretty H, Hawton K, et al: Lithium in the prevention of suicidal behavior and all-mortality in patients with mood disorders: a systematic review of randomized trials. Am J Psychiatry 2005;162:1805–1819.
32 Lauterbach E, Felber W, Müller-Oerlinghausen B, et al: Adjunctive lithium treatment in the prevention of suicidal behavior in depressive disorders: a randomized, placebo-controlled 1 year trial. Acta Psychiatr Scand 2008;118:469–479.
33 Yerevanian BI, Kock RJ, Mintz J: Lithium, anticonvulsants and suicidal behaviour in bipolar disorder. J Affect Disord 2003;73:223–228.
34 Ahrens B, Müller-Oerlinghausen B: Does lithium exert an independent antisuicidal effect? Pharmacopsychiatry 2001;34:132–136.
35 Bierbrauer J, Nilsson A, Müller-Oerlinghausen B, et al: Therapeutic and prophylactic effects of lithium on pathological aggression; in Bauer M, Grof P, Müller-Oerlinghausen B (eds): Lithium in Neuropsychiatry. London, Informa, 2006, pp 227–236.
36 Goodwin FK: Anticonvulsant therapy and suicide in affective disorders. J Clin Psychiatry 1999;60(suppl 2):89–93.
37 Thies-Fletchner K, Müller-Oerlinghausen B, Seibert W, et al: Effect of prophylactic treatment on suicide risk in patients with affective disorders: data from a randomized prospective study. Pharmacopsychiatry 1996;29:103–107.
38 Goodwin FK: Rationale for using lithium in combination with other mood stabilizers in the management of bipolar disorder. J Clin Psychiatry 2003;64:18–24.
39 Sondergard L, Lopez AG, Andersen PK, et al: Mood-stabilizing pharmacological treatment in bipolar disorders and risk of suicide. Bipolar Disord 2008;10:87–94.
40 Collins JC, McFarland BH: Divalproex, lithium and suicide among Medicaid patients with bipolar disorder. J Affect Disord 2008;107:23–28.
41 Joffe R: Does lithium save lives? J Psychiatry Neurosci 2004;29:9.
42 Licht RW, Vestergaard P, Brodersen A: Long-term outcome of patients with bipolar disorder commenced on lithium during hospitalization: a complete 15-year register-based follow-up. Bipolar Disord 2008;10:79–86.
43 Soares-Weiser K, Bravo Vergel V, Revnun S, et al: A systematic review and economic model of the clinical effectiveness and cost-effectiveness of interventions for preventing relapse in people with bipolar disorder. Health Technol Assess 2007;11:206.
44 Cade JF: Lithium salts in the treatment of psychotic excitement. Med J Aust 1949;36:349–352.
45 Forssman H, Walinder J: Lithium treatment of atypical indications. Acta Psychiatr Scand Suppl 1969;207:34–40.
46 Baastrup P: Practical clinical viewpoints regarding treatment with lithium. Acta Psychiatr Scand Suppl 1969;207:12–18.
47 Campbell M, Small AM, Green WH, et al: Behavioral efficacy of haloperidol and lithium carbonate. A comparison in hospitalized aggressive children with conduct disorder. Arch Gen Psychiatry 1984;41:650–656.
48 Campbell M, Adams PB, Small AM, et al: Lithium in hospitalized children with conduct disorder: a double-blind and placebo controlled study. J Am Acad Child Adolesc Psychiatry 1995;34:445–453.
49 Rifkin A, Karajgi B, Dicker R, et al: Lithium treatment of conduct disorders in adolescents. Am J Psychiatry 1997;154:554–555.
50 Malone RP, Bennett DS, Luebbert JF, et al: Aggression classification and treatment response. Psychopharmacol Bull 1998;34:41–45.
51 Carlson GA, Rapport MD, Pataki CS, et al: Lithium in hospitalized children at 4 and 8 weeks: mood, behavior and cognitive effects. J Child Psychol Psychiatry 1992;33:411–425.
52 Malone RP, Delaney MA, Luebbert JF, et al: A double-blind placebo-controlled study of lithium in hospitalized aggressive children and adolescents with conduct disorder. Arch Gen Psychiatry 2000;57:649–654.
53 Dostal T, Zvolsky P: Antiaggressive effect of lithium salts in severe mentally retarded adolescents. Int Pharmacopsychiatry 1970;5:203–207.
54 Micev V, Lynch DM: Effect of lithium on disturbed severely mentally retarded patients. Br J Psychiatry 1974;125:111.
55 Luchins DJ, Dojka D: Lithium and propanolol in aggression and self-injurious behaviour in the mentally retarded. Psychopharmacol Bull 1989;25:372–375.
56 Glenn MB, Wroblewski B, Parziale J, et al: Lithium carbonate for aggressive behaviour or affective instability in ten brain-injured patients. Am J Phys Med Rehabil 1989;68:221–226.
57 Bellus SB, Stewart D, Vergo JG, et al: The use of lithium in the treatment of aggressive behaviours with two brain-injured individuals in a state psychiatric hospital. Brain Inj 1996;10:849–860.
58 Van Putten T, Sanders DG: Lithium in treatment failures. J Nerv Ment Dis 1975;161:255–264.

59 Rifkin A, Quitkin F, Carrillo C, et al: Lithium carbonate in emotionally unstable character disorder. Arch Gen Psychiatry 1972;27:519–524.
60 Prado-Lima P, Knijnik L, Juruena M, et al: Lithium reduces maternal child abuse behavior: a preliminary report. J Clin Pharm Ther 2001;26:279–282.
61 Dorrego MF, Canevaro L, Kuzis G, et al: A randomized, double-blind, crossover study of methylphenidate and lithium in adults with attention-deficit/hyperactivity disorder: preliminary findings. J Neuropsychiatry Clin Neurosci 2002;14:289–295.
62 Sheard MH: Effect of lithium in human aggression. Nature 1971;230:113–114.
63 Tupin JP, Smith DB, Clanon TL, et al: The long-term use of lithium in aggressive prisoners. Compr Psychiatry 1973;14:311–317.
64 Sheard MH: Lithium in the treatment of aggression. J Nerv Ment Dis 1975;160:108–118.
65 Sheard MH, Marini JL, Bridges CI, et al: The effect of lithium on impulsive aggressive behaviour in man. Am J Psychiatry 1976;133:1409–1413.
66 McElroy SL, Pope HG Jr, Keck PE Jr, et al: Are impulse-control disorders related to bipolar disorder? Compr Psychiatry 1996;37:229–240.
67 Müller-Oerlinghausen B: Lithium long-term treatment – does it act via serotonin? Pharmacopsychiatry 1985;18:214–217.
68 Coccaro EF: Central serotonin and impulsive aggression. Br J Psychiatry 1989;8(suppl):52–62.
69 Coccaro EF, Siever LJ, Klar HM, et al: Serotonergic studies in affective and personality disorder patients; correlates with suicidal and impulsive aggressive behavior. Arch Gen Psychiatry 1989;46:587–599.
70 Coccaro EF, Kavoussi RJ, Sheline YI, et al: Impulsive aggression in personality disorder correlates with tritiated paroxetine binding in the platelet. Arch Gen Psychiatry 1996;53:531–536.
71 Roggenbach J, Müller-Oerlinghausen B, Franke L, et al: Peripheral serotonergic markers in acutely suicidal patients. 1. Comparison of serotonergic platelet measures between suicidal individuals, nonsuicidal patients with major depression and healthy subjects. J Neural Transm 2007;114:479–478.

Neuropsychobiology

Neuropsychobiology 2010;62:50–60
DOI: 10.1159/000314310

Published online: May 7, 2010

Novel Insights into Lithium's Mechanism of Action: Neurotrophic and Neuroprotective Effects

Jorge A. Quiroz[a] Rodrigo Machado-Vieira[b] Carlos A. Zarate, Jr.[b] Husseini K. Manji[c]

[a]Hoffman-La Roche Inc., Pharma Development and Exploratory Neuroscience, Nutley, N.J.,
[b]Experimental Therapeutics, Mood and Anxiety Disorders Research Program, NIMH-NIH, Bethesda, Md., and
[c]Johnson & Johnson Pharmaceutical Research and Development, L.L.C., Titusville, N.J., USA

Key Words
Neurodegenerative disorders · B-cell lymphoma 2 · Protein kinase C · Glycogen synthase kinase 3 · Brain-derived neurotrophic factor · Arachidonic acid · Bipolar disorder

Abstract
The monovalent cation lithium partially exerts its effects by activating neurotrophic and neuroprotective cellular cascades. Here, we discuss the effects of lithium on oxidative stress, programmed cell death (apoptosis), inflammation, glial dysfunction, neurotrophic factor functioning, excitotoxicity, and mitochondrial stability. In particular, we review evidence demonstrating the action of lithium on cyclic adenosine monophosphate (cAMP)-mediated signal transduction, cAMP response element binding activation, increased expression of brain-derived neurotrophic factor, the phosphatidylinositide cascade, protein kinase C inhibition, glycogen synthase kinase 3 inhibition, and B-cell lymphoma 2 expression. Notably, we also review data from clinical studies demonstrating neurotrophic effects of lithium. We expect that a better understanding of the clinically relevant pathophysiological targets of lithium will lead to improved treatments for those who suffer from mood as well as neurodegenerative disorders.

Over the past several years, significant evidence has expanded our understanding of how lithium might exert its mood-stabilizing properties in individuals suffering from bipolar disorder. As a result of novel insights into these mechanisms, recent work has demonstrated that this monovalent cation induces its cellular and molecular effects, at least partially, by activating neurotrophic and neuroprotective pathways and its associated signaling mechanisms. Although the changes that lithium exerts to produce mood-stabilizing effects have not been completely clarified, a growing body of evidence supports that neurotrophic cascades might be the common denominator underlying lithium's therapeutic efficacy. In this review, we evaluate each of the currently identified mechanisms of action, as well as the comparatively limited evidence available from human studies demonstrating the neurotrophic and neuroprotective effects of lithium.

Neurotrophic Signaling Cascades

Neurotrophins (NTs) are a family of regulatory factors. They are known to mediate the differentiation and survival of neurons, as well as the modulation of synaptic

KARGER

Fax +41 61 306 12 34
E-Mail karger@karger.ch
www.karger.com

Accessible online at:
www.karger.com/nps

Husseini K. Manji, MD, FRCPC
Johnson & Johnson Pharmaceuticals Group
1125 Trenton-Harbourton Road, E32000
Titusville, NJ 08560 (USA)
Tel. +1 609 730 2968, Fax +1 609 730 2940, E-Mail hmanji@its.jnj.com

transmission and synaptic plasticity. Members of the NT family include nerve growth factor (NGF), brain-derived neurotrophic factor (BDNF), NT-3, NT-4, NT-5, and NT-6. BDNF and other neurotrophic factors are necessary for the survival and function of neurons; thus, sustained reductions of these factors could affect neuronal viability.

The acute effects of BDNF on synaptic plasticity and neurotransmitter release include the release of glutamate, γ-aminobutyric acid, dopamine, and serotonin [1] although it is perhaps best known for its long-term neurotrophic and neuroprotective effects, which may be key to its putative role in the pathophysiology and treatment of mood disorders. Endogenous neurotrophic factors have traditionally been viewed as increasing cell survival by providing necessary trophic support; however, it is now clear that their survival-promoting effects are largely mediated by inhibiting cell death (apoptosis) cascades [1]. Increasing evidence suggests that neurotrophic factors inhibit cell death cascades by activating the extracellular-regulated kinase (ERK) signaling pathway [cyclic adenosine monophosphate (cAMP) response element binding (CREB) is directly phosphorylated and activated by phospho-ERK1/2], the phospholipase C (PLC)-γ cascade, and the phosphoinositide 3-kinase (PI3K)/Akt pathway. Complementarily, phospho-CREB reductions observed after chronic stress [2] could subsequently downregulate the transcription of some neurotrophic genes such as B-cell lymphoma 2 (bcl-2) and BDNF (fig. 1).

Enhanced bcl-2 expression can offset the potentially deleterious consequences of stress-induced neuronal endangerment, suggesting that pharmacologically induced upregulation of bcl-2 may be useful in treating a variety of disorders associated with endogenous or acquired impairments of cellular resilience. In this context, it is notable that severe stress exacerbates stroke outcome by suppressing bcl-2 expression [3]; for instance, following ischemia, 70% less bcl-2 mRNA was expressed by mice exposed to aggressive social stress compared with unstressed mice. Furthermore, stress greatly exacerbated the infarct area in control mice, but this effect was not seen in transgenic mice constitutively expressing increased neuronal bcl-2. Similarly, high corticosterone concentrations were significantly correlated with larger infarcts in wild-type mice but not in transgenic mice overexpressing bcl-2. Overall, it is clear that the neurotrophic factor-ERK/mitogen-activated protein (MAP) kinase-bcl-2 signaling cascade plays a critical role in cell survival in the central nervous system (CNS), and that a fine balance is maintained between the levels and activities of cell survival and cell death factors. In parallel, dysregulation of the necessary coordination between ERK, CREB, and BDNF may also be a key mechanism via which prolonged stress induces atrophy of selective subpopulations of vulnerable neurons and/or distal dendrites. Conceivably, the precise kinetics of ERK and CREB activation could ultimately dictate whether the activated kinases participate in cell death- or survival-promoting pathways.

Lithium Modifies cAMP-Mediated Signal Transduction

Lithium has complex effects on cAMP-mediated signaling, mainly by elevating basal adenylyl cyclase (AC) activity, but also by reducing receptor-stimulated responses in both preclinical and clinical studies (reviewed in Jope [4]). G proteins modulate intracellular cAMP levels by mediating the effect of neurotransmitters (via extracellular receptors) on AC (which catalyzes the conversion of adenosine triphosphate to cAMP). Indeed, preclinical studies conducted by several independent investigators found that the ability of the receptor-mediated signal to be propagated via AC is decreased after lithium treatment [4, 5]. These extensive cellular findings are consistent with an animal model in which the cholera toxin induced G_s and G_{olf} protein hyperactivity when injected into the nucleus accumbens of rats; furthermore, cholera toxin-induced hyperactivity was decreased after lithium administration [6], consistent with decreased G_s and/or G_{olf} activity during lithium treatment (fig. 1). However, while stimulated levels were decreased, there was evidence to suggest increased basal cAMP activity [4]. These complex and potentially regional specific effects on basal activity and stimulated AC activity may arise from the effects of lithium on G proteins and AC subtypes, as well as their relative abundance in different brain regions [4].

The physiologic effects of cAMP are primarily mediated by activation of protein kinase A (PKA), an enzyme that phosphorylates and regulates many proteins including ion channels, cytoskeletal elements, transcription factors, and other enzymes. The transcription factor CREB is one of PKA's major direct targets in the CNS (PKA phosphorylates and activates CREB), and it plays a major role in long-term neuroplasticity (although the cAMP signaling pathway does much more than simply regulate CREB activity). As mentioned above, one of the genes activated by CREB is BDNF, a protein implicated in neuronal survival and synaptic plasticity. A growing

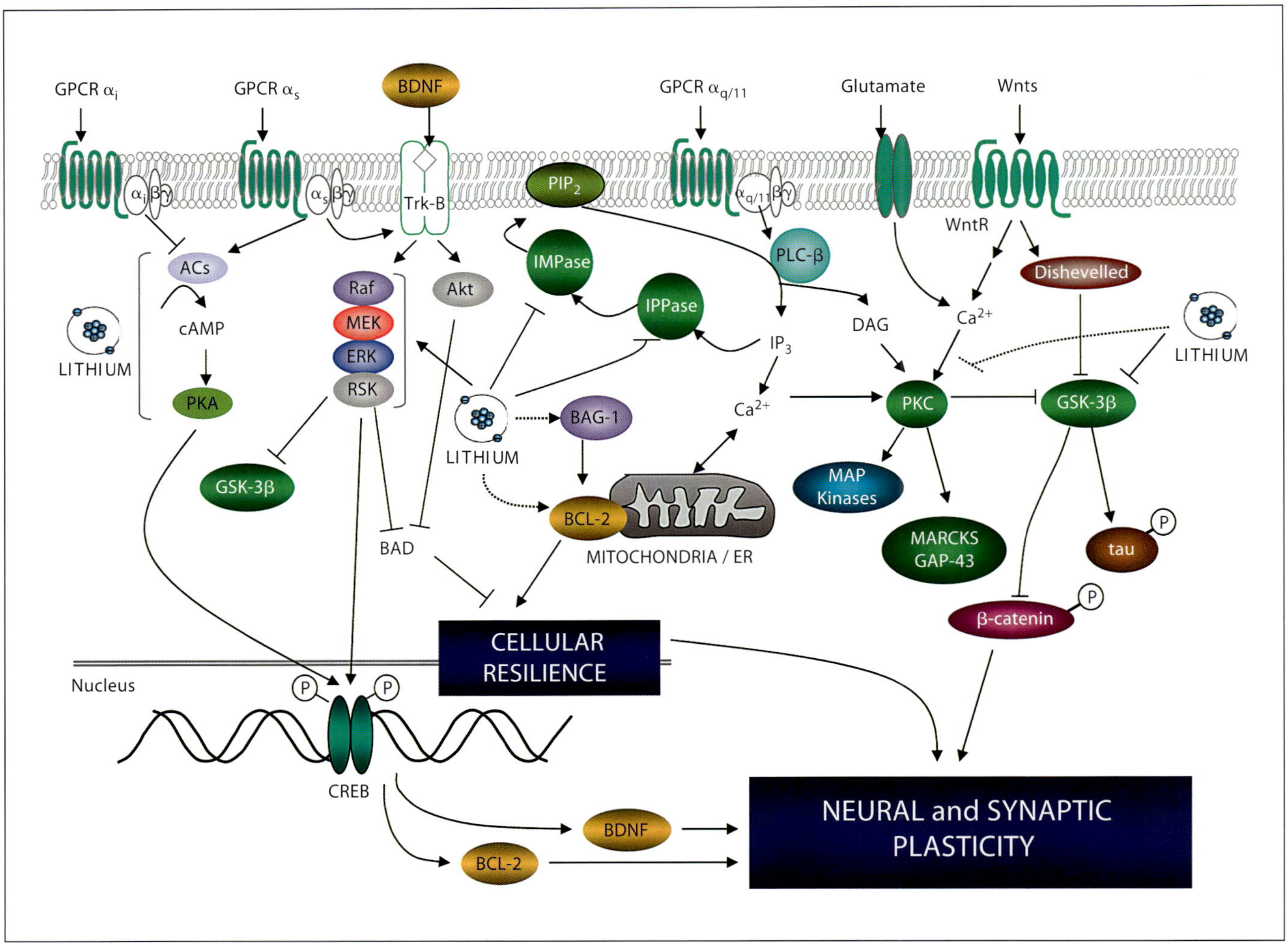

Fig. 1. Neurotrophic and neuroprotective pathways targeted by lithium. BDNF receptor (Trk-B) activation activates the ERK/MAPK pathway, which inhibits GSK-3β (a critical cellular target and effector for diverse proteins) and bad. This activation increases the expression of nuclear CREB, in turn facilitating the expression of neurotrophic/neuroprotective proteins such as bcl-2 and BDNF itself. Mitochondrial bcl-2 also inhibits pro-apoptotic activation of bad, as well as consequent mitochondrial increases of calcium influx and cytochrome c release. Dysregulated intracellular calcium levels, which may increase the risk of cellular apoptosis, have been associated with the pathophysiology of bipolar disorder. Lithium downregulates ER calcium release via an IP_3R-dependent mechanism, and also increases bcl-2 expression, which improves mitochondrial stability and prevents the activation of apoptotic cascades. IMPase, also directly inhibited by lithium, recycles IP_3. In addition, cellular signaling through Wnt glycoproteins and frizzled receptors inhibits GSK-3β. Lithium's inhibition of GSK-3β prevents β-catenin phosphorylation and stimulates its translocation to the nucleus, thus targeting the transcription of specific genes activating neurotrophic effects and synaptogenesis. Different neurotransmitters target receptors coupled to G proteins. Among these, D_1, D_5, and β-adrenergic receptors are coupled to Gα stimulatory proteins that activate AC; H_3, D_2, D_3, and D_4 receptors are coupled to Gα inhibitory proteins that inhibit AC; serotonergic, α_1-adrenergic, M_1, M_3, and M_5 receptors are coupled to $G_{q/11}$, which activates PLC. PLC, in turn, hydrolyses PIP_2 to IP_3 and DAG. DAG activates PKC, which plays a significant role in regulating pre- and postsynaptic aspects of neurotransmission and diverse cellular processes. Lithium also indirectly inhibits PKC. The text provides a complete description of these interactions. Arrows represent 'activation', perpendicular lines represent 'inhibition', and dotted lines represent 'indirect effects'. bad = bcl-2-associated death promoter; bag-1 = bcl-2-associated athanogene; GPCR = G-protein-coupled receptors; IP_3R = IP_3 receptor; MARCKS = myristoylated alanine-rich C kinase substrate; Raf, MEK, ERK, RSK = components of the ERK pathway; Trk-B = tropomyosin receptor kinase.

body of data suggests that agents that directly modulate the cAMP-PKA-CREB-BDNF signaling cascade may be of particular interest for the development of novel agents to treat depressive disorders [7]. Interestingly, lithium and valproate – both of which are mood stabilizers used in the treatment of bipolar disorder – are known to increase BDNF levels in the brains of rats treated chronically with these drugs [8–10].

Lithium Activates CREB and Increases BDNF Expression

At therapeutically relevant concentrations, both lithium and valproate activated the ERK/MAP kinase cascade in human neuroblastoma SH-SY5Y cells in vitro [11]; in vivo, these agents activated the same cascade in the hippocampus and frontal cortex areas of the rodent brain [8]. Lithium also activated ERK1/2 after ischemia, and significantly increased cell proliferation in the hippocampal dentate gyrus [12]. Via CREB, the activation of the ERK/MAP kinase pathway initiates the transcription of BDNF, and also induces bcl-2 gene expression. Consistent with the activation of neurotrophic signaling cascades, chronic treatment of rats with therapeutically equivalent lithium or valproate concentrations increased the activation of ribosomal S6 kinase (a member of the MAPK signaling pathway) and CREB, and eventually doubled bcl-2 levels in the frontal cortex, as evidenced by an increased number of bcl-2 immunoreactive cells in layers II and III of the frontal cortex [13–16].

Several preclinical studies have also investigated the effects of lithium on CREB phosphorylation and activity with mixed overall results [8, 17, 18]. For instance, administration of lithium or valproate increased expression of BDNF in the rodent brain [8, 9], particularly in the hippocampus [19] and frontal cortex [20]. Therapeutic concentrations of lithium selectively increased levels of exon IV-containing BDNF mRNA, and the activity of BDNF promoter IV [21]. While studies have obtained both positive and negative results after lithium exposure [22, 23], most recent evidence suggests that the neurotrophic effect of lithium in cortical neurons requires BDNF expression [24]. In humans, recent data suggested that the Val66Met BDNF gene polymorphism may be associated with the degree of prophylactic response to lithium; there was a trend for the Met allele genotype to be associated with a higher incidence in positive responders to lithium compared to nonresponders [25]. On the other hand, postmortem studies have noted that individuals with bipolar disorder treated with lithium had reduced CREB phosphorylation [26, 27], although the postmortem instability of phosphorylated proteins is a well-known concern in interpreting such reports.

Finally, the modification of other neurotrophic factors has also been associated with chronic lithium treatment. Changes in NGF and glial cell line-derived neurotrophic factor were observed in a rat model of depression [10], including significant increases of NGF concentrations in the frontal cortex, limbic forebrain, hippocampus, and amygdala of adult rats [28]. Lithium also increased serum and hippocampal NT-3 levels in an animal model of mania [29], and upregulated vascular endothelial growth factor in brain endothelial cells and astrocytes [30]. Interestingly, vascular endothelial growth factor has been implicated in neuronal survival, neurotrophic effects, regeneration, growth, and differentiation. Consistent with these effects on neurotrophic signaling cascades, lithium was found to be neuroprotective in other animal and cell models of neuronal insult and disease [31, 32], to promote neurogenesis in the hippocampus of rats, and to increase the regeneration of CNS axons [33].

Lithium Modifies the PI Cascade and Inhibits Protein Kinase C

Inositol phospholipids play a key role in receptor-mediated signal transduction pathways, and are implicated in a variety of responses including cell division, secretion, neuronal excitability, and responsiveness. The PI pathway is initiated by the activation of G-protein-coupled receptors, which pair neurotransmitter receptors to multiple types of intracellular effector proteins. M1, M2, M3, a1, and serotonin receptors coupled to $G\alpha_{q/11}$ induce PLC, which hydrolyzes phosphatidylinositol-4,5-bisphosphate (PIP_2) to yield 2 second messengers: inositol-1,4,5-triphosphate (IP_3) and diacylglycerol (DAG). IP_3 and DAG subsequently modulate the activity of many intracellular events. IP_3 binds to the IP_3 receptor [facilitating the release of calcium from intracellular stores, particularly in the endoplasmic reticulum (ER)] [34, 35], and DAG activates protein kinase C (PKC) (fig. 1).

The direct effect of lithium on inositol monophosphatase (IMPase) [36, 37] and, secondarily, on inositol polyphosphate 1-phosphatase [38, 39] led to the inositol depletion hypothesis of lithium action [40, 41]. This hypothesis suggests that lithium exerts its mood-stabilizing effects by inhibiting IMPase, thereby decreasing inositol concentrations and the amount of PIP_2 available for signaling cascades that rely upon this pathway; this includes, but is not limited to, the neurotrophin signaling pathways, the receptor tyrosine kinase pathways, and the G-protein-mediated signaling, all of which rely on PI availability [40]. The brain is believed to be particularly sensitive to lithium, due to the relatively poor penetration of inositol across the blood-brain barrier [40] or to a reduced ability of specific neuronal populations to transport inositol across their cell membranes [42]. In fact, it has been shown that lithium, carbamazepine, and valproate inhibit sodium myo-inositol transporter 1 (SMIT1) on astrocyte-like cells at therapeutically relevant concentrations [43]. More recently, homozygote knockout mice for the SMIT1 gene (receiving inositol supplementation) appeared to behave similarly to lithium-treated animals in seizure and depression models [44]. However, it has also been reported that reduced intracellular inositol in the brains of SMIT1 knockout mice had no effect on PI levels [45], suggesting that inositol depletion may not have major effects on PI-mediated signaling in this paradigm. Importantly, it has also been reported that the content of SMIT1 mRNA in neutrophils of untreated patients with bipolar I disorder is higher than in a control population while the levels are reduced in treated versus control and untreated bipolar subjects [46]. Although there are some limitations associated with the peripheral nature of the cells studied (that may not reflect brain pathophysiology) and the post hoc origin of the analysis, this hypothesis generating research confirms the need to better understand the role that PI pathways play in bipolar disorder.

As mentioned above, lithium interacts with the PI/PKC pathway by inhibiting IMPase, resulting in decreased free myo-inositol and the subsequent production of DAG, with the downstream effect of also decreasing PKC levels and activity (in some cell culture models, there is evidence of a biphasic action showing initial PKC activation followed by downregulation). PKC is an omnipresent enzyme, and highly enriched in the brain, where it regulates both pre- and postsynaptic aspects of neurotransmission [47], as well as several cellular processes. These include the stimulation of transmembrane glucose transport, secretion, exocytosis, smooth muscle contraction, gene expression, modulation of ion conductance, cell proliferation, and desensitization of extracellular receptors [47]. Interestingly, certain PKC isoforms phosphorylate and inactivate glycogen synthase kinase 3 (GSK-3) in vitro [48] (see below).

PKC isoforms differ in their structure, subcellular localization, tissue specificity, mode of activation, and substrate specificity. It is important to note that chronic lithium treatment decreases the level of PKC isozymes α and ε [49–51] in cells as well as in treated rodents (in part due to the ability of lithium to inhibit IMPase [32, 49]). Following chronic treatment in rats, lithium also decreases the levels and phosphorylation of myristoylated alanine-rich C kinase substrate, a major PKC substrate that has been implicated in signaling and neuroplastic events associated with cytoskeletal architecture [52, 53].

Taken together with the abundant preclinical biochemical and behavioral data supporting the notion that PKC activity may mediate manic-like behaviors, the inhibition of PKC by lithium led to a series of hypothesis-driven clinical studies investigating this relationship. The first study of a fairly selective PKC inhibitor in humans was a small, open-label trial of tamoxifen, which was found to produce a greater than 50% decrease in manic symptoms in 5 of 7 subjects [54]. Furthermore, this effect was very recently confirmed in 2 double-blind, placebo-controlled studies of acutely manic patients by the original researchers [55], and by an independent group of investigators [56].

Action of Lithium on the Arachidonic Acid Signaling Cascade

Lithium also plays an important role in the arachidonic acid (AA) cascade. AA is an important mediator of second messenger pathways in the brain [57, 58], and is released from membrane phospholipids via the activation of receptor/G-protein-initiated phospholipase A2 (PLA2) [59]. This results in AA release from the cellular membrane, and cyclooxygenase-mediated production of eicosanoid metabolites such as prostaglandins and thromboxanes. Due to their lipid-permeable nature, these metabolites mediate numerous subsequent intracellular responses as well as transynaptic responses. In rats, treatment with lithium or valproate resulted in selective reductions in the turnover rate in the brain phospholipids of AA [60–62]. In the case of lithium, an 80% reduction of AA turnover was observed. In addition, lithium decreased the gene expression and protein levels of an AA-specific PLA2 (specifically, cytosolic PLA2) [63, 64], as well as cyclooxygenase 2 protein levels [65]. These findings suggest that the effects of mood stabilizers on cell membranes – and specifically on AA turnover – might be relevant to the pharmacological action of mood stabilizers [58, 62].

Competition of Lithium with Magnesium

Lithium also inhibits some enzymes through direct competition with magnesium, an often-required cofactor [5, 66, 67]. At least 4 related phosphomonoesterases are significantly inhibited at therapeutic serum lithium concentrations [68] (0.6–1.2 mM); in mammals, this group of magnesium-dependent, lithium-sensitive phosphatases includes inositol polyphosphate 1-phosphatase and IMPase (discussed above), fructose 1,6-bisphosphatase, bisphosphate nucleotidase [69], and phosphoglucomutase [70–73]. A significant amount of research has focused on IMPase as a possible therapeutically relevant target of lithium inhibition, predominantly due to the role this enzyme plays in CNS functions [74].

GSK-3 Inhibition by Lithium

GSK-3 is a serine/threonine kinase that regulates diverse cellular processes and directly regulates cell apoptosis. It is key to glycogen synthesis, gene transcription, synaptic plasticity, apoptosis (cell death), cellular structure and resilience, and the circadian cycle [75], all of which are significantly implicated in the pathophysiology of severe recurrent mood disorders. In 1996, GSK-3 was identified as the lithium target responsible for developmental effects in Xenopus embryos [76], but more recently, further evidence substantially supports the claim that GSK-3 is one of the *therapeutic* targets of lithium.

GSK-3β activation functionally inhibits CREB, β-catenin (an important component of memory consolidation), and other survival-promoting transcription factors. GSK-3 is also directly regulated by signals originating from a number of different signaling pathways including the Wnt pathway, the PI3K pathway, PKA, and PKC. Its other targets include transcription factors like c-Jun, proteins bound to microtubules (Tau, microtubule-associated protein 1B, kinesin light chain), cell cycle mediators (cyclin D), and metabolic regulators (glycogen synthase, pyruvate dehydrogenase) [77]. GSK-3 also directly regulates the dopaminergic, glutamatergic, and serotonergic neurotransmitter systems (reviewed in Beaulieu et al. [78] and Jope and Roh [79]). It is interesting to note that increases in phosphorylated GSK-3 levels have also been observed by 5-HT1A receptor activation, 5-HT2 receptor blockage [80], and by atypical antipsychotic administration (with D_2 but also 5-HT2A blocking activity) in experiments with mice [81].

Early studies suggested that peripheral administration of lithium inhibited GSK-3 in the brains of 7-day-old rats [82], as well as during long-term treatment at therapeutic concentrations. Investigators also found that 9 days of lithium treatment in rats (at a mean serum concentration of 0.8 mM) increased cytosolic protein levels of β-catenin, a transcription factor directly regulated by GSK-3 [83]. Small but significant decreases in β-catenin mRNA levels (reflecting cellular compensation) accompanied this increase, further suggesting that lithium exerted its actions post-translationally by inhibiting GSK-3 [83]. Similar findings have noted that chronic lithium does indeed activate β-catenin-dependent transcription in the mouse brain [84] (fig. 1).

Because GSK-3 inhibition is commonly associated with the neurotrophic effects of different survival factors, this kinase may mediate the neuroprotective effects of lithium. In fact, GSK-3 inhibition directly influences gene transcription, leading to anti-apoptotic effects and improved cell structural stability [85]. Notably, diverse studies have noted that GSK-3 is downregulated by lithium, thereby inducing direct neuronal protection against different injuries [86], and providing new insights into the neurotrophic effects of lithium (reviewed in Jope [4]). Furthermore, recent evidence suggests that the behavioral effects of lithium, at least in rodent models, may also be due to GSK-3 inhibition. For instance, the administration of GSK-3 inhibitors resulted in antidepressant-like effects in the forced swim test paradigm following either peripheral lithium administration in rats [71], or lithium administration in mice [84], including intracerebral ventricle injections in mice [87]. While initial reports suggested that knocking out a single copy of the GSK-3β gene in mice resulted in antidepressant-like effects analogous to lithium administration [84], these findings were subsequently not confirmed in mice of a different genetic background [88].

More recently, transgenic mice overexpressing β-catenin demonstrated changes comparable to those observed following lithium administration. These changes included decreased immobility time in the forced swim test and the inhibition of D-amphetamine-induced hyperlocomotion. Such findings are consistent with the notion that the behavioral effects of lithium are mediated via direct inhibition of GSK-3 and the consequent increase in β-catenin [89].

Effect of Lithium on bcl-2 and Mitochondrial Function

The regulatory effects of lithium on apoptosis-controlling proteins appear to occur in both the mitochondria [35] and the ER [90–92]. Relatedly, altered calcium dynamics are the most reproducible biological measure in the pathophysiology of bipolar disorder (reviewed in Quiroz et al. [35] and Warsh et al. [93]). A large movement of calcium into the mitochondria will exceed the mitochondrial capacity to export protons, potentially interrupting adenosine triphosphate synthesis and the activation of the permeability transition pore with release of cytochrome c, thus initiating cellular apoptosis [35]. In addition, excessive production of reactive oxygen species (or free radicals) triggered by mitochondrial dysfunction may lead to oxidative stress, regardless of whether or not this is related to lower antioxidant capacity.

Mitochondria are well known for their critical role in regulating energy production via oxidative phosphorylation, regulation of intracellular calcium, and as critical mediators of cellular apoptosis. Increasing evidence suggests that they may also be integrally involved in general processes of synaptic plasticity [35]. As noted previously, the bcl-2 family of proteins includes both pro- and anti-apoptotic proteins embedded in the inner mitochondrial membrane, although they may also be present in nuclear membranes and in the ER. Therefore, expression and/or activation of pro-apoptotic bcl-2 family members (e.g. bad and bax) may increase mitochondrial membrane permeability, while anti-apoptotic members (e.g. bcl-2 and bcl-xl) have the opposite effect [35]. Interestingly, lithium also increases the expression of bcl-2-associated athanogene (bag-1), which is known to attenuate glucocorticoid receptor nuclear translocation, to activate the ERK/MAP kinases, and to potentiate the anti-apoptotic functions of bcl-2 [94] (fig. 1).

Several biochemical changes have been hypothesized to account for the neurotrophic properties of lithium against oxidative stress and apoptosis, including its ability to regulate bcl-2. bcl-2 levels were found to be robustly increased in the frontal cortex after lithium treatment, particularly in layers II and III [13]. Also, long-term lithium treatment of cultured cerebellar granule cells induced concentration-dependent decreases in p53 mRNA levels as well as bax protein levels (both of which are pro-apoptotic), while acute treatment had no effect. Conversely, bcl-2 mRNA and protein levels increased considerably with long-term lithium treatment, and the bcl-2/bax protein level ratio increased approximately 5-fold after lithium treatment for 5–7 days [16]. Furthermore, chronic treatment with lithium in drinking water prevented the aluminum-induced translocation of cytochrome c, upregulating bcl-2 and bcl-xl, and thus reducing DNA damage. As noted previously, animal models have also noted a relationship between stress and bcl-2 expression.

Human Studies Demonstrating the Neurotrophic Effects of Lithium

While the preclinical data demonstrating that lithium has neurotrophic and neuroprotective effects are striking, considerable caution must be exercised in extrapolating these data to human clinical situations, where the data regarding the neurotrophic effects of lithium are considerably more limited.

The most replicated finding from structural neuroimaging studies is the association between lithium treatment and increased gray matter (GM) volume in brain areas implicated in emotional processing and cognitive control, including the anterior cingulate gyrus, amygdala, and hippocampus [95, 96]. In structural studies, reanalyzed data demonstrated approximately 40% reductions in subgenual prefrontal cortex volumes in individuals with familial mood disorders [97]. These investigators also studied glial cell densities in a small number of individuals with major depressive disorder and found that they exhibited reduced glial cell densities; in contrast, in individuals with bipolar disorder, only those who had discontinued chronic treatment with lithium or valproate exhibited similar reductions [98], suggesting that these mood stabilizers conferred neuroprotective properties.

More recently, a longitudinal high-resolution volumetric MRI study of well-characterized, medication-free individuals with bipolar depression compared total brain GM, prefrontal GM, and left subgenual GM at baseline and after 4 weeks of blinded lithium treatment [93]. Significant increases in total brain GM were observed after chronic lithium administration. Furthermore, region-specific analyses revealed significant differences between lithium responders (>50% decrease in Hamilton Depression Rating Scale scores) and nonresponders; only the lithium responders had significant increases in GM volume in the prefrontal cortex and a trend level increase in the left subgenual prefrontal cortex volume [99]. Another study found that individuals with bipolar disorder not treated with lithium had significantly reduced left anterior cingulate volumes compared to healthy volunteers and lithium-treated patients [100]. Additional MRI studies compared GM and white matter volumes in untreated and lithium-treated patients with bipolar disorder and healthy controls, and found that total GM volume was significantly higher in individuals treated with lithium than in untreated patients and healthy controls [101]. GM density was also found to be significantly greater in diffuse cortical regions in individuals with bipolar disorder who were treated with lithium relative to healthy controls [102]. Interestingly, total hippocampal volume was also significantly greater in lithium-treated patients with bipolar disorder compared with healthy controls or unmedicated patients (by 10 and 14%, respectively) [103], also shown in both short-term (1–8 weeks) and long-term lithium treatment in drug-naïve bipolar subjects [104, 105].

Future Directions: The Neurotrophic Effects of Lithium Might Be Relevant in the Treatment of Other Neuropsychiatric Disorders

As the evidence reviewed above makes clear, lithium has garnered considerable interest as a neuroprotective drug. Data from many in vitro models have evaluated the properties of lithium against excitotoxicity or apoptosis, including glutamate-induced excitotoxicity, C_2-ceramide-induced apoptosis, oxygen and glucose deprivation, aluminum, low potassium medium, ouabain-induced LDH, b-bungarotoxin, and pilocarpine-induced mossy fiber sprouting (reviewed in Bowley et al. [98]). This evidence has prompted a considerable amount of research in an effort to clarify the pathophysiological role of the neuroprotective properties of lithium in animal models of several human diseases. These include models of brain ischemia, injury (especially spinal cord and optic nerve injury), infection, irradiation, neurodegeneration and neuroinflammation, Alzheimer's disease, Huntington's disease, amyotrophic lateral sclerosis, HIV-associated cognitive impairments, spinocerebellar ataxia, and cranial irradiation (reviewed in Machado-Vieira et al. [106]). Interestingly, lithium also appears to stimulate the biogenesis of mitochondria in the CNS and spinal cord, thus inducing neurogenesis and neuronal differentiation also in subjects with amyotrophic lateral sclerosis [107]. Promising preliminary results were obtained from these preclinical studies, leading to current epidemiological and clinical studies addressing the putative use of lithium in human populations. While the results are still to come, these are expected to significantly improve our understanding of these devastating diseases.

Conclusions

Recent studies investigating the robust neurotrophic and neuroprotective effects of lithium have identified novel, exciting, and promising targets. Due to the accumulating data linking lithium mechanisms of action with the enhancement of neurotrophic cascades and pathways, we are optimistic that this evidence will ultimately lead to a better understanding of clinically relevant pathophysiological targets, and the consequent development of improved treatments for those who suffer from these devastating psychiatric and neurodegenerative disorders.

Acknowledgement

We would like to acknowledge the outstanding editorial assistance of Ioline Henter, MA.

References

1 Du J, Gould TD, Manji HK: Neurotrophic signaling in mood disorders; in Finkel T, Gutkind JS (eds): Signal Transduction and Human Disease. Hoboken, Wiley, 2003, pp 411–446.
2 Trentani A, Kuipers SD, Ter Horst GJ, Den Boer JA: Selective chronic stress-induced in vivo erk1/2 hyperphosphorylation in medial prefrontocortical dendrites: implications for stress-related cortical pathology? Eur J Neurosci 2002;15:1681–1691.
3 DeVries AC, Joh HD, Bernard O, Hattori K, Hurn PD, Traystman RJ, Alkayed NJ: Social stress exacerbates stroke outcome by suppressing bcl-2 expression. Proc Natl Acad Sci USA 2001;98:11824–11828.
4 Jope RS: A bimodal model of the mechanism of action of lithium. Mol Psychiatry 1999;4:21–25.
5 Gould TD, Chen G, Manji HK: Mood stabilizer psychopharmacology. Clin Neurosci Res 2002;2:193–212.
6 Kofman O, Li PP, Warsh JJ: Lithium, but not carbamazepine, potentiates hyperactivity induced by intra-accumbens cholera toxin. Pharmacol Biochem Behav 1998;59:191–200.
7 Manji HK, Duman RS: Impairments of neuroplasticity and cellular resilience in severe mood disorders: implications for the development of novel therapeutics. Psychopharmacol Bull 2001;35:5–49.

8 Einat H, Yuan P, Gould TD, Li J, Du J, Zhang L, Manji HK, Chen G: The role of the extracellular signal-regulated kinase signaling pathway in mood modulation. J Neurosci 2003;23:7311–7316.
9 Fukumoto T, Morinobu S, Okamoto Y, Kagaya A, Yamawaki S: Chronic lithium treatment increases the expression of brain-derived neurotrophic factor in the rat brain. Psychopharmacology (Berl) 2001;158:100–106.
10 Angelucci F, Aloe L, Jimenez-Vasquez P, Mathe AA: Lithium treatment alters brain concentrations of nerve growth factor, brain-derived neurotrophic factor and glial cell line-derived neurotrophic factor in a rat model of depression. Int J Neuropsychopharmacol 2003;6:225–231.
11 Yuan PX, Huang LD, Jiang YM, Gutkind JS, Manji HK, Chen G: The mood stabilizer valproic acid activates mitogen-activated protein kinases and promotes neurite growth. J Biol Chem 2001;276:31674–31683.
12 Yan XB, Hou HL, Wu LM, Liu J, Zhou JN: Lithium regulates hippocampal neurogenesis by erk pathway and facilitates recovery of spatial learning and memory in rats after transient global cerebral ischemia. Neuropharmacology 2007;53:487–495.
13 Chen G, Zeng WZ, Yuan PX, Huang LD, Jiang YM, Zhao ZH, Manji HK: The mood-stabilizing agents lithium and valproate robustly increase the levels of the neuroprotective protein bcl-2 in the CNS. J Neurochem 1999;72:879–882.
14 Manji HK, Chen G: Pkc, map kinases and the bcl-2 family of proteins as long-term targets for mood stabilizers. Mol Psychiatry 2002; 7(suppl 1):S46–S56.
15 Chen G, Huang LD, Zeng WZ, Manji HK: Mood stabilizers regulate cytoprotective and mRNA-binding proteins in the brain: long-term effects on cell survival and transcript stability. Int J Neuropsychopharmacol 2001; 4:47–64.
16 Chen RW, Chuang DM: Long-term lithium treatment suppresses p53 and bax expression but increases bcl-2 expression. A prominent role in neuroprotection against excitotoxicity. J Biol Chem 1999;274:6039–6042.
17 Ozaki N, Chuang DM: Lithium increases transcription factor binding to AP-1 and cyclic AMP-responsive element in cultured neurons and rat brain. J Neurochem 1997;69: 2336–2344.
18 Tardito D, Tiraboschi E, Kasahara J, Racagni G, Popoli M: Reduced CREB phosphorylation after chronic lithium treatment is associated with down-regulation of CaM kinase IV in rat hippocampus. Int J Neuropsychopharmacol 2007;10:491–496.
19 Frey BN, Andreazza AC, Rosa AR, Martins MR, Valvassori SS, Reus GZ, Hatch JP, Quevedo J, Kapczinski F: Lithium increases nerve growth factor levels in the rat hippocampus in an animal model of mania. Behav Pharmacol 2006;17:311–318.
20 Jacobsen JP, Mork A: The effect of escitalopram, desipramine, electroconvulsive seizures and lithium on brain-derived neurotrophic factor mRNA and protein expression in the rat brain and the correlation to 5-HT and 5-HIAA levels. Brain Res 2004;1024: 183–192.
21 Yasuda S, Liang MH, Marinova Z, Yahyavi A, Chuang DM: The mood stabilizers lithium and valproate selectively activate the promoter IV of brain-derived neurotrophic factor in neurons. Mol Psychiatry 2009;14: 51–59.
22 McQuillin A, Rizig M, Gurling HM: A microarray gene expression study of the molecular pharmacology of lithium carbonate on mouse brain mRNA to understand the neurobiology of mood stabilization and treatment of bipolar affective disorder. Pharmacogenet Genomics 2007;17:605–617.
23 Hammonds MD, Shim SS, Feng P, Calabrese JR: Effects of subchronic lithium treatment on levels of BDNF, Bcl-2 and phospho-CREB in the rat hippocampus. Basic Clin Pharmacol Toxicol 2007;100:356–359.
24 Hashimoto R, Takei N, Shimazu K, Christ L, Lu B, Chuang DM: Lithium induces brain-derived neurotrophic factor and activates TrkB in rodent cortical neurons: an essential step for neuroprotection against glutamate excitotoxicity. Neuropharmacology 2002;43: 1173–1179.
25 Rybakowski JK, Suwalska A, Skibinska M, Szczepankiewicz A, Leszczynska-Rodziewicz A, Permoda A, Czerski PM, Hauser J: Prophylactic lithium response and polymorphism of the brain-derived neurotrophic factor gene. Pharmacopsychiatry 2005;38: 166–170.
26 Stewart RJ, Chen B, Dowlatshahi D, MacQueen GM, Young LT: Abnormalities in the cAMP signaling pathway in post-mortem brain tissue from the Stanley Neuropathology Consortium. Brain Res Bull 2001;55: 625–629.
27 Young LT, Bezchlibnyk YB, Chen B, Wang JF, MacQueen GM: Amygdala cyclic adenosine monophosphate response element binding protein phosphorylation in patients with mood disorders: effects of diagnosis, suicide, and drug treatment. Biol Psychiatry 2004;55: 570–577.
28 Hellweg R, Lang UE, Nagel M, Baumgartner A: Subchronic treatment with lithium increases nerve growth factor content in distinct brain regions of adult rats. Mol Psychiatry 2002;7:604–608.
29 Walz JC, Frey BN, Andreazza AC, Cereser KM, Cacilhas AA, Valvassori SS, Quevedo J, Kapczinski F: Effects of lithium and valproate on serum and hippocampal neurotrophin-3 levels in an animal model of mania. J Psychiatr Res 2008;42:416–421.
30 Warner-Schmidt JL, Duman RS: VEGF as a potential target for therapeutic intervention in depression. Curr Opin Pharmacol 2008;8: 14–19.
31 Chuang DM, Chen R, Chalecka-Franaszek E, Ren M, Hashimoto R, Senatorov V, Kanai H, Hough C, Hiroi T, Leeds P: Neuroprotective effects of lithium in cultured cells and animal models of diseases. Bipolar Disord 2002;4:129–136.
32 Manji HK, Moore GJ, Chen G: Lithium at 50: have the neuroprotective effects of this unique cation been overlooked? Biol Psychiatry 1999;46:929–940.
33 Huang X, Wu DY, Chen G, Manji H, Chen DF: Support of retinal ganglion cell survival and axon regeneration by lithium through a bcl-2-dependent mechanism. Invest Ophthalmol Vis Sci 2003;44:347–354.
34 Gould TD, Manji HK: Signaling networks in the pathophysiology and treatment of mood disorders. J Psychosom Res 2002;53:687–697.
35 Quiroz JA, Gray NA, Kato T, Manji HK: Mitochondrially mediated plasticity in the pathophysiology and treatment of bipolar disorder. Neuropsychopharmacology 2008; 33:2551–2565.
36 Naccarato WF, Ray RE, Wells WW: Biosynthesis of myo-inositol in rat mammary gland. Isolation and properties of the enzymes. Arch Biochem Biophys 1974;164:194–201.
37 Hallcher LM, Sherman WR: The effects of lithium ion and other agents on the activity of myo-inositol-1-phosphatase from bovine brain. J Biol Chem 1980;255:10896–10901.
38 Ragan CI, Watling KJ, Gee NS, Aspley S, Jackson RG, Reid GG, Baker R, Billington DC, Barnaby RJ, Leeson PD: The dephosphorylation of inositol 1,4-bisphosphate to inositol in liver and brain involves two distinct Li+-sensitive enzymes and proceeds via inositol 4-phosphate. Biochem J 1988;249:143–148.
39 Inhorn RC, Majerus PW: Properties of inositol polyphosphate 1-phosphatase. J Biol Chem 1988;263:14559–14565.
40 Berridge MJ, Downes CP, Hanley MR: Neural and developmental actions of lithium: a unifying hypothesis. Cell 1989;59:411–419.
41 Berridge MJ, Downes CP, Hanley MR: Lithium amplifies agonist-dependent phosphatidylinositol responses in brain and salivary glands. Biochem J 1982;206:587–595.
42 Gani D, Downes CP, Batty I, Bramham J: Lithium and myo-inositol homeostasis. Biochim Biophys Acta 1993;1177:253–269.
43 Lubrich B, van Calker D: Inhibition of the high affinity myo-inositol transport system: a common mechanism of action of antibipolar drugs? Neuropsychopharmacology 1999; 21:519–529.
44 Bersudsky Y, Shaldubina A, Agam G, Berry GT, Belmaker RH: Homozygote inositol transporter knockout mice show a lithium-like phenotype. Bipolar Disord 2008;10: 453–459.
45 Berry GT, Buccafusca R, Greer JJ, Eccleston E: Phosphoinositide deficiency due to inositol depletion is not a mechanism of lithium action in brain. Mol Genet Metab 2004;82: 87–92.

46 Willmroth F, Drieling T, Lamla U, Marcushen M, Wark HJ, van Calker D: Sodium-myo-inositol co-transporter (SMIT-1) mRNA is increased in neutrophils of patients with bipolar 1 disorder and down-regulated under treatment with mood stabilizers. Int J Neuropsychopharmacol 2007;10:63–71.
47 Nishizuka Y: Intracellular signaling by hydrolysis of phospholipids and activation of protein kinase C. Science 1992;258:607–614.
48 Goode N, Hughes K, Woodgett JR, Parker PJ: Differential regulation of glycogen synthase kinase-3 beta by protein kinase c isotypes. J Biol Chem 1992;267:16878–16882.
49 Manji HK, Bersudsky Y, Chen G, Belmaker RH, Potter WZ: Modulation of protein kinase C isozymes and substrates by lithium: the role of myo-inositol. Neuropsychopharmacology 1996;15:370–381.
50 Manji HK, Etcheberrigaray R, Chen G, Olds JL: Lithium decreases membrane-associated protein kinase C in hippocampus: selectivity for the alpha isozyme. J Neurochem 1993;61:2303–2310.
51 Li X, Jope RS: Selective inhibition of the expression of signal transduction proteins by lithium in nerve growth factor-differentiated PC12 cells. J Neurochem 1995;65:2500–2508.
52 Lenox RH, Watson DG, Patel J, Ellis J: Chronic lithium administration alters a prominent PKC substrate in rat hippocampus. Brain Res 1992;570:333–340.
53 Watson DG, Lenox RH: Chronic lithium-induced down-regulation of MARCKS in immortalized hippocampal cells: potentiation by muscarinic receptor activation. J Neurochem 1996;67:767–777.
54 Bebchuk JM, Arfken CL, Dolan-Manji S, Murphy J, Hasanat K, Manji HK: A preliminary investigation of a protein kinase C inhibitor in the treatment of acute mania. Arch Gen Psychiatry 2000;57:95–97.
55 Zarate CA, Singh JB, Carlson PJ, Quiroz JA, Jolkovsky L, Luckenbaugh DA, Manji HK: Efficacy of a protein kinase C inhibitor (tamoxifen) in the treatment of acute mania: a pilot study. Bipolar Disord 2007;9:561–570.
56 Yildiz A, Guleryuz S, Ankerst DP, Ongur D, Renshaw PF: Protein kinase C inhibition in the treatment of mania: a double-blind, placebo-controlled trial of tamoxifen. Arch Gen Psychiatry 2008;65:255–263.
57 Axelrod J, Burch RM, Jelsema CL: Receptor-mediated activation of phospholipase A2 via GTP-binding proteins: arachidonic acid and its metabolites as second messengers. Trends Neurosci 1988;11:117–123.
58 Rapoport SI: In vivo fatty acid incorporation into brain phosholipids in relation to plasma availability, signal transduction and membrane remodeling. J Mol Neurosci 2001;16:243–261; discussion 279–284.
59 Axelrod J: Phospholipase A2 and G proteins. Trends Neurosci 1995;18:64–65.
60 Chang MC, Contreras MA, Rosenberger TA, Rintala JJ, Bell JM, Rapoport SI: Chronic valproate treatment decreases the in vivo turnover of arachidonic acid in brain phospholipids: a possible common effect of mood stabilizers. J Neurochem 2001;77:796–803.
61 Chang MC, Grange E, Rabin O, Bell JM, Allen DD, Rapoport SI: Lithium decreases turnover of arachidonate in several brain phospholipids. Neurosci Lett 1996;220:171–174.
62 Rapoport SI, Bosetti F: Do lithium and anticonvulsants target the brain arachidonic acid cascade in bipolar disorder? Arch Gen Psychiatry 2002;59:592–596.
63 Chang MC, Jones CR: Chronic lithium treatment decreases brain phospholipase A2 activity. Neurochem Res 1998;23:887–892.
64 Rintala J, Seemann R, Chandrasekaran K, Rosenberger TA, Chang L, Contreras MA, Rapoport SI, Chang MC: 85 kDa cytosolic phospholipase A2 is a target for chronic lithium in rat brain. Neuroreport 1999;10:3887–3890.
65 Bosetti F, Rintala J, Seemann R, Rosenberger TA, Contreras MA, Rapoport SI, Chang MC: Chronic lithium downregulates cyclooxygenase-2 activity and prostaglandin E(2) concentration in rat brain. Mol Psychiatry 2002;7:845–850.
66 Amari L, Layden B, Rong Q, Geraldes CF, Mota de Freitas D: Comparison of fluorescence, (31)P NMR, and (7)Li NMR spectroscopic methods for investigating Li(+)/Mg(2+) competition for biomolecules. Anal Biochem 1999;272:1–7.
67 Ryves WJ, Harwood AJ: Lithium inhibits glycogen synthase kinase-3 by competition for magnesium. Biochem Biophys Res Commun 2001;280:720–725.
68 Davies SP, Reddy H, Caivano M, Cohen P: Specificity and mechanism of action of some commonly used protein kinase inhibitors. Biochem J 2000;351:95–105.
69 York JD, Ponder JW, Majerus PW: Definition of a metal-dependent/Li(+)-inhibited phosphomonoesterase protein family based upon a conserved three-dimensional core structure. Proc Natl Acad Sci USA 1995;92:5149–5153.
70 Nordenberg J, Kaplansky M, Beery E, Klein S, Beitner R: Effects of lithium on the activities of phosphofructokinase and phosphoglucomutase and on glucose-1,6-diphosphate levels in rat muscles, brain and liver. Biochem Pharmacol 1982;31:1025–1031.
71 Rhyu GI, Ray WJ Jr, Markley JL: Enzyme-bound intermediates in the conversion of glucose 1-phosphate to glucose 6-phosphate by phosphoglucomutase. Phosphorus NMR studies. Biochemistry 1984;23:252–260.
72 Ray WJ Jr, Szymanki ES, Ng L: The binding of lithium and of anionic metabolites to phosphoglucomutase. Biochim Biophys Acta 1978;522:434–442.
73 Masuda CA, Xavier MA, Mattos KA, Galina A, Montero-Lomeli M: Phosphoglucomutase is an in vivo lithium target in yeast. J Biol Chem 2001;276:37794–37801.
74 Gould TD, Einat H, Bhat R, Manji HK: AR-A014418, a selective GSK-3 inhibitor, produces antidepressant-like effects in the forced swim test. Int J Neuropsychopharmacol 2004;7:387–390.
75 Jope RS: Lithium and GSK-3: one inhibitor, two inhibitory actions, multiple outcomes. Trends Pharmacol Sci 2003;24:441–443.
76 Klein PS, Melton DA: A molecular mechanism for the effect of lithium on development. Proc Natl Acad Sci USA 1996;93:8455–8459.
77 Gould TD, Zarate CA, Manji HK: Glycogen synthase kinase-3: a target for novel bipolar disorder treatments. J Clin Psychiatry 2004;65:10–21.
78 Beaulieu JM, Gainetdinov RR, Caron MG: Akt/GSK3 signaling in the action of psychotropic drugs. Annu Rev Pharmacol Toxicol 2009;49:327–347.
79 Jope RS, Roh MS: Glycogen synthase kinase-3 (GSK3) in psychiatric diseases and therapeutic interventions. Curr Drug Targets 2006;7:1421–1434.
80 Li X, Bijur GN, Jope RS: Glycogen synthase kinase-3beta, mood stabilizers, and neuroprotection. Bipolar Disord 2002;4:137–144.
81 Li X, Rosborough KM, Friedman AB, Zhu W, Roth KA: Regulation of mouse brain glycogen synthase kinase-3 by atypical antipsychotics. Int J Neuropsychopharmacol 2007;10:7–19.
82 Munoz-Montano JR, Moreno FJ, Avila J, Diaz-Nido J: Lithium inhibits Alzheimer's disease-like tau protein phosphorylation in neurons. FEBS Lett 1997;411:183–188.
83 Gould TD, Chen G, Manji HK: In vivo evidence in the brain for lithium inhibition of glycogen synthase kinase-3. Neuropsychopharmacology 2004;29:32–38.
84 O'Brien WT, Harper AD, Jove F, Woodgett JR, Maretto S, Piccolo S, Klein PS: Glycogen synthase kinase-3beta haploinsufficiency mimics the behavioral and molecular effects of lithium. J Neurosci 2004;24:6791–6798.
85 Chin PC, Majdzadeh N, D'Mello SR: Inhibition of GSK3beta is a common event in neuroprotection by different survival factors. Brain Res Mol Brain Res 2005;137:193–201.
86 Bijur GN, De Sarno P, Jope RS: Glycogen synthase kinase-3beta facilitates staurosporine- and heat shock-induced apoptosis. Protection by lithium. J Biol Chem 2000;275:7583–7590.
87 Kaidanovich-Beilin O, Milman A, Weizman A, Pick CG, Eldar-Finkelman H: Rapid antidepressive-like activity of specific glycogen synthase kinase-3 inhibitor and its effect on beta-catenin in mouse hippocampus. Biol Psychiatry 2004;55:781–784.

88 Bersudsky Y, Shaldubina A, Kozlovsky N, Woodgett JR, Agam G, Belmaker RH: Glycogen synthase kinase-3beta heterozygote knockout mice as a model of findings in postmortem schizophrenia brain or as a model of behaviors mimicking lithium action: negative results. Behav Pharmacol 2008;19:217–224.

89 Gould TD, Einat H, O'Donnell KC, Picchini AM, Schloesser RJ, Manji HK: Beta-catenin overexpression in the mouse brain phenocopies lithium-sensitive behaviors. Neuropsychopharmacology 2007;32:2173–2183.

90 Ghribi O, Herman MM, Spaulding NK, Savory J: Lithium inhibits aluminum-induced apoptosis in rabbit hippocampus, by preventing cytochrome c translocation, Bcl-2 decrease, Bax elevation and caspase-3 activation. J Neurochem 2002;82:137–145.

91 Leng Y, Liang MH, Ren M, Marinova Z, Leeds P, Chuang DM: Synergistic neuroprotective effects of lithium and valproic acid or other histone deacetylase inhibitors in neurons: roles of glycogen synthase kinase-3 inhibition. J Neurosci 2008;28:2576–2588.

92 Yeste M, Alvira D, Verdaguer E, Tajes M, Folch J, Rimbau V, Pallas M, Camins A: Evaluation of acute antiapoptotic effects of Li+ in neuronal cell cultures. J Neural Transm 2007;114:405–416.

93 Warsh JJ, Andreopoulos S, Li PP: Role of intracellular calcium signaling in the pathophysiology and pharmacotherapy of bipolar disorder: current status. Clin Neurosci Res 2004;4:201–213.

94 Zhou R, Gray NA, Yuan P, Li X, Chen J, Chen G, Damschroder-Williams P, Du J, Zhang L, Manji HK: The anti-apoptotic, glucocorticoid receptor cochaperone protein BAG-1 is a long-term target for the actions of mood stabilizers. J Neurosci 2005;25:4493–4502.

95 Moore GJ, Bebchuk JM, Hasanat K, Chen G, Seraji-Bozorgzad N, Wilds IB, Faulk MW, Koch S, Glitz DA, Jolkovsky L, Manji HK: Lithium increases N-acetyl-aspartate in the human brain: in vivo evidence in support of bcl-2's neurotrophic effects? Biol Psychiatry 2000;48:1–8.

96 Phillips ML, Travis MJ, Fagiolini A, Kupfer D: Medication effects in neuroimaging studies of bipolar disorder. Am J Psychiatry 2008;165:313–320.

97 Drevets WC, Price JL, Simpson JR Jr, Todd RD, Reich T, Vannier M, Raichle ME: Subgenual prefrontal cortex abnormalities in mood disorders. Nature 1997;386:824–827.

98 Bowley MP, Drevets WC, Ongur D, Price JL: Low glial numbers in the amygdala in major depressive disorder. Biol Psychiatry 2002;52:404–412.

99 Moore GJ, Cortese BM, Glitz DA, Zajac-Benitez C, Quiroz JA, Uhde TW, Drevets WC, Manji HK: A longitudinal study of the effects of lithium treatment on prefrontal and subgenual prefrontal gray matter volume in treatment-responsive bipolar disorder patients. J Clin Psychiatry 2009;70:699–705.

100 Sassi RB, Brambilla P, Hatch JP, Nicoletti MA, Mallinger AG, Frank E, Kupfer DJ, Keshavan MS, Soares JC: Reduced left anterior cingulate volumes in untreated bipolar patients. Biol Psychiatry 2004;56:467–475.

101 Sassi RB, Nicoletti M, Brambilla P, Mallinger AG, Frank E, Kupfer DJ, Keshavan MS, Soares JC: Increased gray matter volume in lithium-treated bipolar disorder patients. Neurosci Lett 2002;329:243–245.

102 Bearden CE, Thompson PM, Dalwani M, Hayashi KM, Lee AD, Nicoletti M, Trakhtenbroit M, Glahn DC, Brambilla P, Sassi RB, Mallinger AG, Frank E, Kupfer DJ, Soares JC: Greater cortical gray matter density in lithium-treated patients with bipolar disorder. Biol Psychiatry 2007;62:7–16.

103 Bearden CE, Thompson PM, Dutton RA, Frey BN, Peluso MA, Nicoletti M, Dierschke N, Hayashi KM, Klunder AD, Glahn DC, Brambilla P, Sassi RB, Mallinger AG, Soares JC: Three-dimensional mapping of hippocampal anatomy in unmedicated and lithium-treated patients with bipolar disorder. Neuropsychopharmacology 2008;33:1229–1238.

104 Yucel K, McKinnon MC, Taylor VH, Macdonald K, Alda M, Young LT, MacQueen GM: Bilateral hippocampal volume increases after long-term lithium treatment in patients with bipolar disorder: a longitudinal MRI study. Psychopharmacology (Berl) 2007;195:357–367.

105 Yucel K, Taylor VH, McKinnon MC, Macdonald K, Alda M, Young LT, MacQueen GM: Bilateral hippocampal volume increase in patients with bipolar disorder and short-term lithium treatment. Neuropsychopharmacology 2008;33:361–367.

106 Machado-Vieira R, Manji HK, Zarate CA Jr: The role of lithium in the treatment of bipolar disorder: convergent evidence for neurotrophic effects as a unifying hypothesis, Bipolar Disord 2009;11(suppl 2):92–109.

107 Fornai F, Longone P, Ferrucci M, Lenzi P, Isidoro C, Ruggieri S, Paparelli A: Autophagy and amyotrophic lateral sclerosis: the multiple roles of lithium. Autophagy 2008;4:527–530.

Neuropsychobiology

Neuropsychobiology 2010;62:61–71
DOI: 10.1159/000314311

Published online: May 7, 2010

Pharmacogenetics of Lithium Long-Term Treatment: Focus on Initiation and Adaptation Mechanisms

Alessandro Serretti Antonio Drago

Institute of Psychiatry, University of Bologna, Bologna, Italy

Key Words

Lithium · Pharmacogenetics · Bipolar disorder · Manic depression · Gene

Abstract

Bipolar disorder is a common disease with a high impact in terms of personal suffering and socioeconomic burden. The disentanglement of the molecular deregulations that cause this disorder is pivotal to the understanding of its etiology. This will hopefully cast the engineering of new and more favorable treatments. New insights in the molecular aspects of bipolar disorder may be brought by the understanding of the pharmacodynamics of lithium, the first-line treatment for this disease. The mechanisms by which lithium exerts its activity in the central nervous system are not fully clarified: it is hypothesized that lithium may drive acute molecular events whose activation over time triggers long-lasting modifications in critical neuronal nets. These events are associated with long-lasting changes in the expression profile of genes in neurons that are embedded in these crucial nets. The molecular events that are acutely and chronically triggered by lithium will be reviewed here and matched with the evidence that arises from the pharmacogenetics investigations. Moreover, the pharmacogenetics reports that are not strictly associated with the mechanisms that are thought to be acutely and chronically elicited by lithium will be included in the final part of the paper.

Introduction

Bipolar disorder (BD) is a common disease which chronically exposes patients to switches between manic and depressive states [1]. The lifetime prevalence of BD in 2000 was 3.7% in the United States, and 1% in the UK, Germany and Italy [2], even though underdiagnosis is a concrete concern and the real prevalence may likely be higher [2, 3]. Besides the high costs in terms of personal sufferance, it is estimated that BD caused a loss of GBP 1,510 million in the UK for unemployment and GBP 179 million for suicide per year during the period 1999–2000 [2]. Lithium is an effective treatment for BD [4, 5]: understanding the pharmacodynamics of lithium could delineate the putative key disruptions related to the pathophysiology of BD [6]. The pharmacodynamics of lithium may be separated according to the targets of its activity: first-line targets may be considered as the enzymes that are immediately inhibited at the therapeutic concentra-

KARGER

Fax +41 61 306 12 34
E-Mail karger@karger.ch
www.karger.com

0302–282X/10/0621–0061$26.00/0

Accessible online at:
www.karger.com/nps

Alessandro Serretti, MD, PhD
Institute of Psychiatry, University of Bologna
Viale Carlo Pepoli 5
IT–40123 Bologna (Italy)
Tel. +39 051 658 4233, Fax +39 051 521 030, E-Mail alessandro.serretti@unibo.it

tion of lithium (acute effect); second-line targets may be considered as enzymes that are not directly inhibited by lithium but whose activity varies if first-line targets are continuously inhibited (intermediate effect); third-line targets may be considered as those which cause a more permanent change in the neuronal activity as a consequence of the events occurring at first- and second-line targets (chronic effect). This model refers to the 'initiation and adaptation' paradigm [7] which postulates that some molecular events are initiated after acute exposure to treatment and their ongoing activation determines chronic adaptations which restore the physiological orchestration of the neuronal nets and are then associated with the healing properties of the treatment. In the following paragraphs, the evidence of the pharmacogenetics of lithium will be described in the context of the 'initiation and adaptation' paradigm. The acute events will be treated first. A final paragraph will be dedicated to the evidence which could not easily be reduced to this paradigm, this perhaps being dependent on the incomplete knowledge of molecular mechanisms of neurons and glia and neural nets' interplay.

Acute Events after Lithium Exposure

The Acute Molecular Mechanisms

At a therapeutic concentration, lithium acutely inhibits the following enzymes: inositol monophosphatase (isoenzymes 1 and 2); inositol polyphosphate 1-phosphatase; bisphosphate nucleotidase; fructose 1,6-bisphosphastase; phosphoglucomutase, and glycogen synthase kinase-3β (GSK-3β) [8]. The enzymes associated with the inositol pathway and GSK-3β are of particular interest: the inositol depletion theory and the theory based on the activity of the enzyme GSK-3β are well-known models for the pharmacodynamics of lithium. Inositol is a cellular metabolite whose transformations convey a vast array of information through the increase in the intracellular Ca^{2+} concentration and the activation of protein kinase C (PKC). Metabolism, secretion, phototransduction, cell growth and differentiation are associated with the inositol pathway [9–14]. Inositol is supplied and stored in cells in a number of metabolic ways: (1) glucose 6 phosphate metabolism, (2) Ins(1,4,5)P3 metabolism, (3) glycerophospholipid metabolism and (4) environment. Lithium inhibits these paths in an uncompetitive way [15–17], which is intuitively consistent with the therapeutic action of the cation: it is proposed that lithium specifically downregulates the upregulated neuronal pathways in the prefrontal cortex which may be the cause of BD [15, 18]. This line of thought is referred to as the 'inositol depletion theory'. Even though this paradigm is suggestive and consistent, there is some conflicting evidence: other substances which inhibit the inositol pathway (bisphosphonates, terpene and tropolone) [19] are not as effective as lithium in clinical practice. Moreover, animal and imaging studies provided nonunivocal evidence supporting the 'inositol depletion theory' [20–26]. GSK-3β is inhibited by lithium at a therapeutic concentration: it is a highly conserved kinase, originally identified as a regulator of glycogen synthesis but involved in a vast array of metabolic paths (table 1).

Challenging the Acute Molecular Events: Results from Pharmacogenetics Studies

Dimitrova et al. [27] reported no significant association in a set of variations of IMPA2 (–461C→T/rs2075824; rs3786282; 599 + 97G→A; 599 + 99G→A) with lithium response. The sample under study consisted of 121 Bulgarian bipolar patients and all their parents, 116 parent-offspring trios and 174 unrelated BP patients, matched for age and gender with 170 healthy blood donors. An interesting study investigating both IMPA1 and IMPA2 in 2 different samples (Norwegian and Palestinian Arab trios) found a positive association in the Arab sample (–461C→T and –207T→C) only for IMPA2 investigated variations [28] while no evidence supporting a role of IMPA1 on bipolar phenotype or response to lithium was found. This strengthens the hypothesis that lithium-sensitive or bipolar risk-associated inositol monophosphatase activity is more related to the IMPA2 genetic sequence than to IMPA1. Anyway, more studies are warranted to define this path in detail, and the gene has not been investigated completely yet. The genetic association analyses conducted on GSK-3β partially supported its putative relevance as a mediator of the BD phenotype and of lithium efficacy. Our group first reported that a variation located in the promoter of the GSK-3β (–50T/C) was associated with some clinical features in a bipolar sample such as age at onset and response to total sleep deprivation. In some details, 60 depressed bipolar type I inpatients were treated with the mood stabilizer and it was found that homozygotes for the mutant allele (C) of the GSK-3β promoter showed a later onset of illness and better acute effects of total sleep deprivation treatment on perceived mood. The limited frequency of the C/C genotype in the sample did not allow a final conclusion, even though that finding stood for a certain influence of GSK-3β in BD phenotypes. Consistently, the association be-

Table 1. Some of the most relevant molecular paths that are under the GSK-3 control (http://www.genome.jp/kegg/kegg2.html)

Glycogen synthase	Glycogen metabolism	Phosphorylated and thereby inhibited by GSK-3 mainly after a prior phosphorylation performed by a different kinase referred to as casein kinase II. This is quite relevant because the lack of inhibition of the glycogenosynthesis in neurons leads to neuronal death
Wnt	A signaling pathway associated with many cellular activities; 4 out of them are of particular interest: (1) Wnt/β-catenin; (2) planar cell polarity pathway; (3) PKC, and (4) Ca^{2+}/protein kinase A pathway	Phosphorylated and thereby inhibited by GSK-3 mainly after a prior phosphorylation performed by a different kinase *Wnt/β-catenin:* β-catenin can enter the nucleus and promotes the expression of a set of genes. When activated, GSK-3β constitutively phosphorylates β-catenin and promotes its ubiquination and degradation *Planar cell polarity pathway:* also inhibited by GSK-3; the details of this pathway are scarcely known *PKC:* it mediates a wide array of cellular activities including ERK path, NF-κB (transcription factors) and CPI (involved in the smooth muscle contractile elements) pathways *Ca^{2+}/protein kinase A pathway:* relies on the activation of phospholipase C and the production of DAG and inositol triphosphate
Tau	Tau proteins are microtubule-associated proteins that are abundant in neurons in the central nervous system. Tau proteins interact with tubulin to stabilize microtubules. Phosphorylation of tau results in disruption from microtubule organization and it is associated with the pathophysiology of Alzheimer's disease	Tau protein hyperphosphorylation can be mediated by SSK-3, increased β-amyloid production and inflammatory responses and may then be associated with the neurodevelopmental disorder which may underlie BD
Erb signaling	A cellular path which is associated with the activity of neuroregulins and epidermal growth factors	Activation of ErbB starts the PKB/Akt cascade which is associated with diverse cell events, including the inhibition of GSK-3 activity and the activation of the MAPK cascade
Hedgehog	This signaling system is associated with the transduction of information which is necessary for an embryo for proper development. If corrupted it can result in malformations and diseases such as cancer. Associated with the regeneration following nervous system injury and disease	Involved in cell fate determination, pattern formation, proliferation, and differentiation in multiple tissue types
Akt kinase activity	The serine/threonine kinase Akt mediates mitogenic and anti-apoptotic responses that result from activation of multiple signaling cascades	The integration of the intracellular signals with Akt regulates the phosphorylation of several downstream effectors, such as nuclear factor B (NF-B), the mammalian target of rapamycin (mTOR), Forkhead, Bad and GSK-3. It has been found to be an alternative way of signaling transduction after dopamine binding to D_2 class (D_2, D_3, D_4 associated with Gαi and reduced production of cAMP). Akt system is also involved in the phosphorylation of GSK in the insulin signaling pathway
Axon guidance	Involved in neurodevelopment and neuronal cross talking	GSK-3-mediated phosphorylation is associated with a positive regulation of CRMP (collapsin response mediator protein) activity along with the action of phosphatidylinositol 3-kinase, protein kinase B and collapsing. CRMP proteins are mediators of Sema3A (semaphorin 3A) signaling and neuronal differentiation
Focal adhesion	A key event in cell migration and definition of cell nets	FAK (focal adhesion kinase) is a key regulating enzyme in this path
Melanogenesis	Melanin received some attention in the psychiatric field as plasma-soluble melanins can form oxidative polymerization and copolymerization of dopa, catecholamines and other substances which can concentrate in tissues and cause an oxidative damage. Moreover, reactive oxygen side products (ROSP) are generated during and after melanogenesis	GSK-3 can impact the melanogenesis by the inhibition of the β-catenin cascade
CycD	Cyclin D is part of the CycD-Cdk4 complex and it promotes cell growth	It is one of the targets of GSK-3

tween the –50T/C genotype and age at onset was replicated shortly thereafter in a wider sample (n = 185) of bipolar patients [29, 30], and 1 year later we reported that the same polymorphism was associated with better prophylaxis in a sample of 88 bipolar patients treated with lithium and whose recurrence rate was detected 2 years before and 2 years during the lithium treatment [31]. Indeed, these results are not definitive: the first study we performed was underpowered to detect a little genetic effect, and poor replication evidence is reported so far. Consistently, some replication findings reported negative association results [32]: the study by Michelon et al. [32] did not consider the –50T/C variation, but focused on –1727A/T. Szczepankiewicz et al. [33] also could not support the suggestions we reported about the pharmacogenetics of lithium, even though they found a positive association between the –50T/C variation and bipolar II disorder in female patients [34].

Intermediate Events after Lithium Exposure

The Intermediate Molecular Mechanisms

The 'initiation and adaptation' paradigm postulates that chronic adaptation occurs after the ongoing stimulation of the early targets of lithium. It has been reported that the main targets of the acute effect of lithium are the enzymes associated with the metabolism of inositol and GSK-3β. The molecular paths that are associated with the activity of these enzymes are quite numerous: recent studies helped the identification of the most relevant paths which are considered as the putative adapted mechanisms that underline the activity of lithium. Two out of them are related to the activity of PKC and to the activity of the mitogen-activated protein kinases (MAPK). This result is referred to a recent notable experiment that has been conducted by Seelan et al. [35]: the changes in the expression profile of a set of 44,000 genes during a 33-day-long lithium exposure were measured in a population of human neuronal cells. As a result, 671 differentially regulated transcripts were identified, 347 out of which were upregulated, and the others were found to be downregulated. Interestingly, the analysis of the expression profile of the different genes revealed that key modulated targets after lithium exposure were PKC and MAPK. PKC exists as a family of closely related subspecies and has a heterogeneous distribution in the brain. It is involved in a vast array of activities and takes part in the regulation of neuronal excitement, neurotransmitter release and long-term alterations in gene expression and neural plasticity [36–38]. It has been reported that chronic lithium treatment results in an isoenzyme-specific decrease in PKC-α and PKC-η in the frontal cortex and hippocampus [39, 40], and that hyperactivity of PKC in the prefrontal cortex is associated with impairment of neuronal function [41]. PKC is associated with the signaling that results from the stimulation of relevant receptors such as noradrenergic α_1, serotonergic 2A and metabotropic glutamate. These receptors are associated with the activation of the G protein Gq which is followed by the activation of phospholipase C which in turn cleaves PIP2 to form inositol triphosphate and diacylglycerol (DAG). Inositol triphosphate activates some receptors on the surface of the endoplasmic reticulum to release Ca^{2+} which facilitates the translocation of PKC to the cell membrane where it is activated by DAG. An enzyme referred to as DAG kinase inhibits this pathway, phosphorylating DAG and starting its metabolization back to PIP2. Another relevant actor in this pathway is the regulator of G protein signaling 4 which inhibits Gq. The regulator of G protein signaling 4 and DAG kinase have been found to be altered in mental diseases [41]. Finally, PKC interacts with the MAPK path which regulates cell survival, and one of its targets is the myristoylated alanine-rich C kinase substrate which is relevant to neuroplastic events and whose regulation has been demonstrated to be impacted by chronic lithium treatment [42]. On the other hand, MAPK activity is related to the transmission of extracellular signals to the nucleus triggering the expression of specific genes. In particular, MAPK has been associated with the activity of relevant endogenous growth factors such as nerve growth factor and brain-derived neurotrophic factor (BDNF). MAPK exerts its activity via 3 main pathways in mammalian cells: (1) extracellular signal-regulated kinases, (2) c-Jun NH2 terminal kinases and (3) p38 [43].

Challenging the Intermediate Molecular Events: Results from Pharmacogenetics Studies

It would be worth challenging the hypothesis that PKC and MAPK regulations are fundamental to the antimanic activity of lithium by analyzing the interactions between the genetic variations that are located in the key enzymes of these paths and the response to the cation in humans. Unfortunately, despite the relevance of these modulators, a set of systematic pharmacogenetic association studies on humans treated with lithium is still lacking. There is one interesting exception, a study recently published by Mamdani et al. [62], that will be discussed in the next paragraph. The relevance of the PCK and of the MAPK pathways has been demonstrated in an elegant expression study on animals [35]. Other investigations reported results that are only partially overlapping: Plant et al. [44] described the alteration of the expression profile of a set of genes in a neuronal blastoma line (14 statistically significant changes in expression after lithium exposure) and showed that a protein related to the anti-apoptotic mechanism referred to as Six 1 was found to have a 31.3-fold increased expression after lithium exposure. Six 1 inhibits the apoptotic mechanism through the blockade of caspase 3 activation, thus further stressing the relevance of the neuroprotective activity of lithium. Interestingly, one of the genes whose expression was altered by lithium after 72 h of exposure was found to be protein kinase D1 which belongs to the family of PKC. To further complicate the matter, a recent experiment [45] reported different results from those published by Seelan et al. [35]: administering lithium to mice for 7 days resulted in a different expression profile of 5 genes including metallothionein 3; ATPase, Na/K transporting, α_1-polypeptide; transcription elongation factor B (SIII)-polypeptide 2; proteasome subunit β type 5, and guanine nucleotide-binding protein β_1. These genes are associated with an array of relevant cellular activities which encompass the metal ion homeostasis, the regulation of the RNA polymerase, the regulation of the protein degradation and the transduction of the G-coupled signals. The results by Chetcuti et al. [45] do not overlap those obtained by Seelan et al. [35]. This discrepancy may depend on the different model which has been used: complex organisms respond differently from isolated cells (see, for example, the gephyrin regulation [46]); moreover, 7 days of treatment in mice are roughly comparable to 33 days in humans. Finally, a false-positive bias must be taken into account.

Chronic Events after Lithium Exposure

The Chronic Molecular Mechanisms

It has been reported that the inositol pathway along with the GSK-3β enzyme are the acute targets of lithium treatment, while the cellular events that are associated with the activation of the PKC and of the MAPK pathways are the putatively intermediate targets. Both PKC and MAPK are associated with the information flow triggered in cells after external stimuli, but their activation is ephemeral if not connected with the putatively more persistent changes in the neuronal nets that are triggered by lithium. Identifying the persistent changes associated with the continuously altered activation of PCK and MAPK may be the next relevant step. There is accumulating evidence that these modifications pertain to the factors that modulate the growth and resilience of cells. The cytoprotective protein bcl-2 has been demonstrated to be impacted by chronic lithium treatment [47–49], mainly in the frontal cortex [50], dentate gyrus and striatum [49]. In vivo and in vitro studies confirmed this result [47, 51, 52]. Consistently, lithium was found to reduce the levels of the proapoptotic protein p53 in cellular models [52, 53]. These lines of evidence suggest that lithium has a neuroprotective activity, which has been confirmed in various studies [47, 49]. Interestingly, many growth factors act via induction of the expression of the bcl-2 protein and the inhibition of proapoptotic factors through the MAPK signaling system [54, 55]. Moreover, in the experiment conducted by Seelan et al. [35], the most upregulated gene after chronic treatment with lithium was the one coding for the peroxiredoxin 2, an antioxidant gene. Consistently, the most downregulated protein was found to be the tribbles homolog 3, a proapoptotic protein. The findings by Seelan et al. [35] are of particular interest: a cross search within the public databases revealed that only 7 out of 671 differentially regulated genes had been described before in the literature. This discrepancy opens the way to further research in this direction. Another recent experiment published by Corena-McLeod et al. [56] reported a list of 12 genes whose expression was found to be changed after lithium treatment; the set of proteins under investigation were chosen from synaptosomal preparations. Interestingly, 3 drugs were administered to mice (paliperidone, valproate and lithium): the gene whose expression profile was altered after each treatment was found to be Synapsin 2 (SYN2) (upregulated). SYN1 and SYN2 are involved in mediating the BDNF-mediated modulation of glutamate release [57].

Challenging the Chronic Molecular Events: Results from Pharmacogenetics Studies

Even if not conclusive, pharmacogenetics investigations casted some evidence that variations located in genes associated with growth factors are associated with the prophylactic activity of lithium. In 2005, Rybakowski et al. [58] reported that the Met and the T allele at variations Val66Met and –270C/T located in the BDNF gene were associated with a better response in a sample of 88 bipolar patients observed for a mean period of 15 years (5–27 years). Conflicting with this, in 2006 Michielon et al. [32] found no evidence of an association between lithium prophylactic action and the G196A variation located in the BDNF gene in a sample of 134 bipolar patients all from Caucasian origins. The period of observation was 2 years. The same lack of association was found in the same year by Masui et al. [59] in a sample of 83 bipolar I and 78 bipolar II patients, all of Japanese origins. The period of observation was 1 year in that case. The discrepancy between these findings can be related to the different period of observation which is longer in the study by Rybakowski et al. [58], suggesting that longer periods of observation are needed to detect small genetic impacts. Moreover, the small sample size that characterizes the studies may have seriously dampened their power to detect the small impact determined by a single variation in the phenotype. Interestingly, Rybakowski et al. [60] repeated the study in 2007 in a large sample of patients (n = 111) and found that S individuals (patients with the s/s or s/l genotype at the promoter of the serotonin transporter gene) having the Val/Val genotype at the Val66Met variation located in the BDNF gene were significantly more frequent in nonresponders to the prophylactic activity of lithium compared with responders. The period of observation was the same as in the former study performed by the group [58]. More recently, Dmitrzak-Weglarz et al. [61] reported that the C/G (rs988748) and G/A (rs6265) variations located in the BDNF gene were associated with the prophylactic lithium response in a sample of 108 bipolar patients. In that study, 4 variations located in the BDNF gene were investigated (rs2030324, rs988748, rs6265 (Val66Met) and rs2203877) along with 3 variations located in the functionally related gene NTRK2 (rs1187326, rs2289656, rs1187327). Neither the variations in NTRK2 nor the combination of the variations located in both BDNF and NTRK2 were found to be significantly associated with the prophylactic activity of lithium. Finally, it has recently been reported that 2 variations located in the CREB 1 gene were found to be associated with response to lithium in a sample of 258 patients with BD [62]. These findings support the hypothesis that lithium mediates its antimanic activity via a neurotrophic activity as far as the increased expression of bcl-2, which is associated with the activation of the MAPK pathway, is thought to be mediated by the activation of the cAMP response element-binding protein [63, 64].

Other Targets

The 'initiation and adaptation' paradigm has been adopted in this paper to allocate the most promising candidates associated with the therapeutic efficacy of lithium and with the putative biological disruptions that cause BD. This line of research offers the advantage of a clear-cut rationality but suffers the limits of a quite rigid approach. There is a set of evidence from pharmacogenetic investigations that are not based on this paradigm. For example, mood balance is thought to be dependent on the activity of the monoaminergic systems in the brain [65], a theory which is referred to as the 'monoaminergic theory' [66] and results from the first observations that drugs able to interfere with the monoaminergic tone had an antidepressant activity. Even though the paradigm that results from this theory offers several limits and cannot be accepted as a whole comprehensive explanation of mood disorders, there are several lines of evidence that suggest that the regulators of the monoaminergic systems are involved in the regulation of mood [67, 68]. This given, a set of genes that code for receptors of transporters of the monoaminergic systems have been investigated as possible modulators of the prophylactic activity of lithium. Our group performed some investigations on a set of key genes such as HTR2A, HTR2C, HTR1A [69], DRD2, DRD4, GABRA1 [70], DRD3 [71], COMT, and MAO-A [72] reporting negative association findings. The same lack of association was found in independent studies [73]. This lack of association may be due to an irrelevant role played by these genes in the action of lithium, or it may be related to the limits of these researches: the sample sizes were small and likely underpowered in order to detect the impact of a single gene toward a complex phenotype as the response to the lithium treatment is. The number of patients we enrolled in 1998 were 43 bipolar patients and 12 major depressive patients observed for an average period of 49 months [71]; the next year, we enrolled 100 bipolar patients and 25 patients with major depressive disorder observed for an average period of 53 months [70]. In 2000, we enrolled 100 patients with bipolar disorder and 22 patients with major depressive disorder, and observed them for an average period of 52 months [69]. Two years later, we enrolled 160 bipolar patients and 41 patients suffering from major depressive disorder and observed them for an average period of 59.8 months [72]. The study published by Dmitrzak-Weglarz et al. in 2005 [73] included 92 bipolar patients. Other candidate genes provided more consistent results. The investigation of SERTPR in a sample of 167 bipolar patients observed for a period of 5 years on average showed that the s/s genotype was associated with worse response to lithium treatment [74], and 3 years later we reported that heterozygosis for the same variation was associated with better prophylactic activity of lithium in a sample of 83 bipolar

Table 2. Pharmacogenetic studies on lithium prophylactic action

Author	Gene	Gene variant	Subjects	Length of observation	Results
Mamdani et al. [62]	CREB 1, CREB 2, CREB 3	3 variations located in the promoter; 14 variations located in the UTR regions; 6 variations located in exons, and 15 variations located in introns	Caucasian 180 lithium responders 69 lithium nonresponders 127 controls	more than 13 years	2 variations in the CREB 1 gene were associated with lithium response
Mamdani et al. [88]	PREP	rs3799990; rs3818281; rs1149320; rs6902415; rs12192054; rs1028792; rs9486069; rs9500087; rs720225	Caucasian 249 BPI 126 controls	13.25 ± 7.26 years	no significant association
Michelon et al. [32]	BDNF, INPP1, AP-2β, SERTPR	G196A, C973A, VNTR, s-l	Caucasian 134 BPI	2 years	no significant association
Szczepankiewicz et al. [33]	GSK-3β	T-50C	Caucasian 89 BPI	5 years	no significant association
Masui et al. [89]	BDNF	Val66Met	Japanese 83 BPI 78 BPII	1 year	no significant association
Masui et al. [59]	Xbp-1	–116C/G	Japanese 83 BPI 78 BPII	1 year	C associated with better response
Dmitrzak-Weglarz et al. [73]	HTR2A, HTR2C	T102C, (Cys23Ser)	Caucasian 92 BPI	5 years	no significant association
Benedetti et al. [31]	GSK-3β	–50T/C	Caucasian 88 BPI	more than 2 years before and 2 years on lithium treatment	C associated with better response
Rybakowski et al. [90]	SERTPR	s-l	Caucasians 67 BPI	more than 5 years	S associated with worse response
Rybakowski et al. [58]	BDNF	Val66Met and –270C/T	Caucasians 88 BPI	5–27 years (mean 15 years)	Met and T associated with better response
Rybakowski et al. [60]	BDNF and SERTPR	s-l and Val66Met	Caucasian 111 BPI	5–27 years (mean 15 years)	S individuals having Val/Val genotype were significantly more frequent in nonresponders to the prophylactic activity of lithium compared with responders
Dimitrova et al. [27]	IMPA2	8 SNPs	Caucasian 237 parent-offspring trios 174 BPI 170 controls	3 years	no significant association
Serretti et al. [75]	SERTPR	s-l polymorphism	Caucasian 83 BPI	3 years	l/l associated with poor response l/s associated with better response
Washizuka et al. [91]	mtDNA	10398A/G	Japanese 54 BPI	4.4–5.6 years	A associated with better response
Serretti et al. [72]	COMT, MAO-A, Gβ3	G158A, 30-bp repeat, C825T	Caucasian 160 BPI	4.9 years	no significant association
Serretti et al. [74]	SERTPR	s-l polymorphism	Caucasian 167 BPI	4.85 years	s/s associated with worse response
Serretti et al. [69]	HTR2A, HTR2C, HTR1A	T102C, C-1420T, Cys23Ser, C-1019G	Caucasian 102 BPI	4.3 years	no significant association

Table 2 (continued)

Author	Gene	Gene variant	Subjects	Length of observation	Results
Serretti et al. [76]	TPH	A218C; A779C	Caucasian 90 BPI	4.2 years	TPH*A/A variant showed a trend toward a worse response
Serretti et al. [70]	DRD2, DRD4, GABRA1	SerCys, VNTR, VNTR	Caucasian 100 BPI	4.41 years	no significant association
Steen et al. [92]	INPP1	C973A	Caucasian pilot bipolar sample	–	C973A transversion was present in responders
Serretti et al. [71]	DRD3	BalI	Caucasian 43 BPI	4.08 years	no significant association

UTR = Not coding flanking regions of genes; PREP = prolyl endopeptidase; INPP1 = inositol polyphosphate 1-phosphatase; AP-2 = activator protein-2; SERTPR = promoter of serotonin transporter gene; Xbp-1 = X-box binding protein 1; HTR2A = serotonin receptor 2A; HTR2C = serotonin receptor 2C; IMPA2 = inositol monophosphatase 2; mtDNA = mitochondrial DNA; COMT = catechol-O-methyltransferase; MAO-A = monoamine oxidase A; HTR1A = serotonin receptor 1A; TPH = tryptophan hydroxylase; DRD2 = dopamine receptor D_2; DRD4 = dopamine receptor D_4; GABRA1 = GABA(A) receptor α_1 subunit; DRD3 = dopamine receptor D_3.

patients observed for a period of 3 years [75]. The strength of association was found to be as high as p = 0.005 in the study published in 2004. In 2001, we reported an association as significant as p = 0.009 in a sample of 201 bipolar patients [74]. Our findings were confirmed by the results reported by Rybakowski et al. [60] which have been discussed in the former paragraph. Consistently, we reported that a variation located in the tryptophan hydroxylase gene, an enzyme involved in the metabolism of serotonin, was associated with the prophylactic activity of lithium, even though only a marginal significance was detected [76]. These results are schematically reported in table 2 along with other pharmacogenetics studies that investigated genes that have been poorly reinvestigated thereafter. Some interesting findings have been reported for the glutamate neurotransmission. Interestingly, one of the proteins whose regulation was found to be altered in the experiment by Seelan et al. [35] was GRIA2 (glutamate receptor, ionotropic, downregulated), a finding that replicated the results of Gray et al. [77] who observed a downregulation of synaptosomal and total GluR2 protein in the rat hippocampus after 4 weeks of treatment with lithium. This is of relevance because the disruption of the glutamate balance is thought to be pivotal to the genesis of neurodegenerative diseases such as Alzheimer's and Parkinson's disease, and, putatively, schizophrenia [78–80]. Consistent with this, one of the genes that was found to be overexpressed after lithium treatment in the formerly cited experiment conducted by Plant et al. [44] was the Rho guanine nucleotide exchange factor 12: this enzyme has recently been demonstrated to modulate the vesicular traffic; its overexpression was found to result in enhanced secretion from immature granules, depleting cells of secretory cargos [81]. Even though the refined mechanism by which this depletion that could be triggered by chronic lithium treatment interacts with the monoamine balance or with the neuronal crosstalking in general is still far from being understood, this piece of evidence suggests that lithium activity may be a modulator of the neurotransmitter release.

Conclusion

The mechanisms of action of lithium are still poorly understood. A relevant line of research points toward the direction of the 'initiation and adaptation' paradigm. This theory posits that lithium triggers some events during acute phases (initiation phase) whose activation over time determines chronic adaptations (adaptation phase) that are responsible for the therapeutic activity of the cation. The first targets of lithium are relevant metabolic mediators such as enzymes associated with the metabolism of inositol which mainly activate an array of second messengers including PKC and the increase in the intracellular Ca^{2+} concentration, and GSK-3β, a key enzyme in cell equilibrium which inhibits a set of different molecular paths. There is evidence that the initiation phase is not associated with the healing efficacy of lithium per se as far as substances with the ability to inhibit the inositol path are not as effective as lithium in controlling and preventing manic states. The inhibition of the metabolism of inositol and the inhibition of the activity of the GSK-3 enzyme have been found to be associated with profound adaptations of two main second messenger regulators: the PKC and the MAKP pathway. These adaptations could be able to induce persistent changes in the expression profile of neurons over time. There is a set of evidence that considers the neurotrophic factors as the most relevant adapted mediators after chronic lithium exposure. This is consistent with the ability of lithium to protect cells promoting anti-apoptotic paths, and to inhibit the collapse of sensory neuron growth cones and increase the growth cone area along with other mood stabilizers (valproic acid and carbamazepine) [82]. Intriguingly, this event has been reported to be sensitive to disruption of the inositol orchestration [83]. The efficacy of lithium as a neuroprotective agent makes it a possible candidate for treatment in neurodegenerative disorders such as Alzheimer's disease [84, 85], and it also suggests that BD may be related to a neurodegenerative disorder [86], opening the way to a new definition of this disorder. The lack of replication that characterizes the research on the response to lithium treatment is probably also due to a list of difficulties and bias that dampened the efforts produced so far. The difficulties that are intrinsic to this topic are due to the lack of knowledge about the fine tuning of the regulation of gene expression. Moreover, the definition of the phenotype and the correct classification of bipolar patients is a great challenge to research, especially in the field of lithium genetics [17]. Finally, most of the genetic association studies suffered from flaws that detract from their validity and dampen the rate of replication: we recently reviewed this topic suggesting some methods to better design studies for the genetics applied to depressive disorders, a suggestion that may fit the BD-focused studies as well [87]. Briefly, the specification of the sampling source (inpatients vs. outpatients, primary vs. tertiary settings), the standardization of the treatments, adequate monitoring of the compliance through plasma levels and a sufficient length of observation (from 6 weeks to 3–6 months) and the inclusion of detailed environmental and sociodemographic variables are some suggestions that may be applied to studies that deal with BD patients. Finally, as far as genetics is considered, there are many sources of variations that usually are not taken into account: the complete coverage of the known genetic variations, the inclusion of the copy number variations along with the single nucleotide variations and the measurement of the epigenetic factors such as the methylation rates are not standard so far. The standard methodology should be improved accordingly, in order to obtain more comparable results.

References

1 Kessler RC, Rubinow DR, Holmes C, Abelson JM, Zhao S: The epidemiology of DSM-III-R bipolar I disorder in a general population survey. Psychol Med 1997;27:1079–1089.
2 Fajutrao LB, Locklear JC, Priaulx J, Heyes A: A systematic review of the evidence of the burden of bipolar disorder in Europe. Clin Pract Epidemol Ment Health 2009;5:3.
3 Hirschfeld RM, Calabrese JR, Weissman MM, Reed M, Davies MA, Frye MA, Keck PE Jr, Lewis L, McElroy SL, McNulty JP, Wagner KD: Screening for bipolar disorder in the community. J Clin Psychiatry 2003;64:53–59.
4 Gelenberg AJ, Hopkins HS: Report on efficacy of treatments for bipolar disorder. Psychopharmacol Bull 1993;29:447–456.
5 Geddes JR, Burgess S, Hawton K, Jamison K, Goodwin GM: Long-term lithium therapy for bipolar disorder: systematic review and meta-analysis of randomized controlled trials. Am J Psychiatry 2004;161:217–222.
6 Manji HK, Potter WZ, Lenox RH: Signal transduction pathways. Molecular targets for lithium's actions. Arch Gen Psychiatry 1995;52:531–543.
7 Hyman SE, Nestler EJ: Initiation and adaptation: a paradigm for understanding psychotropic drug action. Am J Psychiatry 1996;153:151–162.
8 Quiroz JA, Gould TD, Manji HK: Molecular effects of lithium. Mol Interv 2004;4:259–272.
9 Seeds AM, York JD: Inositol polyphosphate kinases: regulators of nuclear function. Biochem Soc Symp 2007:183–197.
10 Shears SB: Assessing the omnipotence of inositol hexakisphosphate. Cell Signal 2001;13:151–158.
11 Cockcroft S: Trafficking of phosphatidylinositol by phosphatidylinositol transfer proteins. Biochem Soc Symp 2007:259–271.
12 Tolias KF, Cantley LC: Pathways for phosphoinositide synthesis. Chem Phys Lipids 1999;98:69–77.
13 Irvine RF, Schell MJ: Back in the water: the return of the inositol phosphates. Nat Rev Mol Cell Biol 2001;2:327–338.
14 Resnick AC, Saiardi A: Inositol polyphosphate multikinase: metabolic architect of nuclear inositides. Front Biosci 2008;13:856–866.
15 Berridge MJ, Downes CP, Hanley MR: Neural and developmental actions of lithium: a unifying hypothesis. Cell 1989;59:411–419.
16 Willmroth F, Drieling T, Lamla U, Marcushen M, Wark HJ, van Calker D: Sodium-myo-inositol co-transporter (SMIT-1) mRNA is increased in neutrophils of patients with bipolar 1 disorder and down-regulated under treatment with mood stabilizers. Int J Neuropsychopharmacol 2007;10:63–71.
17 Ikonomov OC, Manji HK: Molecular mechanisms underlying mood stabilization in manic-depressive illness: the phenotype challenge. Am J Psychiatry 1999;156:1506–1514.

18 Kofman O, Belmaker RH: Ziskind-Somerfeld Research Award 1993. Biochemical, behavioral, and clinical studies of the role of inositol in lithium treatment and depression. Biol Psychiatry 1993;34:839–852.
19 Fauroux CM, Freeman S: Inhibitors of inositol monophosphatase. J Enzyme Inhib 1999; 14:97–108.
20 Phiel CJ, Klein PS: Molecular targets of lithium action. Annu Rev Pharmacol Toxicol 2001;41:789–813.
21 Harwood AJ: Lithium and bipolar mood disorder: the inositol-depletion hypothesis revisited. Mol Psychiatry 2005;10:117–126.
22 Harwood AJ, Agam G: Search for a common mechanism of mood stabilizers. Biochem Pharmacol 2003;66:179–189.
23 Agam G, Shamir A, Shaltiel G, Greenberg ML: Myo-inositol-1-phosphate (MIP) synthase: a possible new target for antibipolar drugs. Bipolar Disord 2002;4(suppl 1):15–20.
24 Lenox RH, Hahn CG: Overview of the mechanism of action of lithium in the brain: fifty-year update. J Clin Psychiatry 2000;61(suppl 9):5–15.
25 Lenox RH, Wang L: Molecular basis of lithium action: integration of lithium-responsive signaling and gene expression networks. Mol Psychiatry 2003;8:135–144.
26 Silverstone PH, McGrath BM, Kim H: Bipolar disorder and myo-inositol: a review of the magnetic resonance spectroscopy findings. Bipolar Disord 2005;7:1–10.
27 Dimitrova A, Milanova V, Krastev S, Nikolov I, Toncheva D, Owen MJ, Kirov G: Association study of myo-inositol monophosphatase 2 (IMPA2) polymorphisms with bipolar affective disorder and response to lithium treatment. Pharmacogenomics J 2005;5:35–41.
28 Sjoholt G, Ebstein RP, Lie RT, Berle JO, Mallet J, Deleuze JF, Levinson DF, Laurent C, Mujahed M, Bannoura I, Murad I, Molven A, Steen VM: Examination of IMPA1 and IMPA2 genes in manic-depressive patients: association between IMPA2 promoter polymorphisms and bipolar disorder. Mol Psychiatry 2004;9:621–629.
29 Benedetti F, Serretti A, Colombo C, Lorenzi C, Tubazio V, Smeraldi E: A glycogen synthase kinase 3-beta promoter gene single nucleotide polymorphism is associated with age at onset and response to total sleep deprivation in bipolar depression. Neurosci Lett 2004;368:123–126.
30 Benedetti F, Bernasconi A, Lorenzi C, Pontiggia A, Serretti A, Colombo C, Smeraldi E: A single nucleotide polymorphism in glycogen synthase kinase 3-beta promoter gene influences onset of illness in patients affected by bipolar disorder. Neurosci Lett 2004; 355:37–40.
31 Benedetti F, Serretti A, Pontiggia A, Bernasconi A, Lorenzi C, Colombo C, Smeraldi E: Long-term response to lithium salts in bipolar illness is influenced by the glycogen synthase kinase 3-beta –50 T/C SNP. Neurosci Lett 2005;376:51–55.
32 Michelon L, Meira-Lima I, Cordeiro Q, Miguita K, Breen G, Collier D, Vallada H: Association study of the INPP1, 5HTT, BDNF, AP-2beta and GSK-3beta gene variants and retrospectively scored response to lithium prophylaxis in bipolar disorder. Neurosci Lett 2006;403:288–293.
33 Szczepankiewicz A, Rybakowski JK, Suwalska A, Skibinska M, Leszczynska-Rodziewicz A, Dmitrzak-Weglarz M, Czerski PM, Hauser J: Association study of the glycogen synthase kinase-3beta gene polymorphism with prophylactic lithium response in bipolar patients. World J Biol Psychiatry 2006;7: 158–161.
34 Szczepankiewicz A, Skibinska M, Hauser J, Slopien A, Leszczynska-Rodziewicz A, Kapelski P, Dmitrzak-Weglarz M, Czerski PM, Rybakowski JK: Association analysis of the GSK-3beta T-50C gene polymorphism with schizophrenia and bipolar disorder. Neuropsychobiology 2006;53:51–56.
35 Seelan RS, Khalyfa A, Lakshmanan J, Casanova MF, Parthasarathy RN: Deciphering the lithium transcriptome: microarray profiling of lithium-modulated gene expression in human neuronal cells. Neuroscience 2008; 151:1184–1197.
36 Manji HK, Lenox RH: Long-term action of lithium: a role for transcriptional and post-transcriptional factors regulated by protein kinase C. Synapse 1994;16:11–28.
37 Nishizuka Y: Intracellular signaling by hydrolysis of phospholipids and activation of protein kinase C. Science 1992;258:607–614.
38 Stabel S, Parker PJ: Protein kinase C. Pharmacol Ther 1991;51:71–95.
39 Manji HK, Bebchuk JM, Moore GJ, Glitz D, Hasanat KA, Chen G: Modulation of CNS signal transduction pathways and gene expression by mood-stabilizing agents: therapeutic implications. J Clin Psychiatry 1999; 60(suppl 2):27–39; discussion 40–41, 113–116.
40 Manji HK, Etcheberrigaray R, Chen G, Olds JL: Lithium decreases membrane-associated protein kinase C in hippocampus: selectivity for the alpha isozyme. J Neurochem 1993;61: 2303–2310.
41 Arnsten AF: Ameliorating prefrontal cortical dysfunction in mental illness: inhibition of phosphotidyl inositol-protein kinase C signaling. Psychopharmacology (Berl) 2009; 202:445–455.
42 Lenox RH, McNamara RK, Watterson JM, Watson DG: Myristoylated alanine-rich C kinase substrate (MARCKS): a molecular target for the therapeutic action of mood stabilizers in the brain? J Clin Psychiatry 1996; 57(suppl 13):23–31; discussion 32–33.
43 Gutkind JS: The pathways connecting G protein-coupled receptors to the nucleus through divergent mitogen-activated protein kinase cascades. J Biol Chem 1998;273: 1839–1842.
44 Plant KE, Anderson E, Simecek N, Brown R, Forster S, Spinks J, Toms N, Gibson GG, Lyon J, Plant N: The neuroprotective action of the mood stabilizing drugs lithium chloride and sodium valproate is mediated through the up-regulation of the homeodomain protein Six1. Toxicol Appl Pharmacol 2009;235:124–134.
45 Chetcuti A, Adams LJ, Mitchell PB, Schofield PR: Microarray gene expression profiling of mouse brain mRNA in a model of lithium treatment. Psychiatr Genet 2008;18:64–72.
46 Ikonomov OC, Petrov T, Soden K, Shisheva A, Manji HK: Lithium treatment in ovo: effects on embryonic heart rate, natural death of ciliary ganglion neurons, and brain expression of a highly conserved chicken homolog of human MTG8/ETO. Brain Res Dev Brain Res 2000;123:13–24.
47 Manji HK, Moore GJ, Chen G: Clinical and preclinical evidence for the neurotrophic effects of mood stabilizers: implications for the pathophysiology and treatment of manic-depressive illness. Biol Psychiatry 2000;48: 740–754.
48 Chen G, Huang LD, Jiang YM, Manji HK: The mood-stabilizing agent valproate inhibits the activity of glycogen synthase kinase-3. J Neurochem 1999;72:1327–1330.
49 Manji HK, Moore GJ, Chen G: Lithium at 50: have the neuroprotective effects of this unique cation been overlooked? Biol Psychiatry 1999;46:929–940.
50 Rajkowska G, Halaris A, Selemon LD: Reductions in neuronal and glial density characterize the dorsolateral prefrontal cortex in bipolar disorder. Biol Psychiatry 2001;49: 741–752.
51 Chen G, Rajkowska G, Du F, Seraji-Bozorgzad N, Manji HK: Enhancement of hippocampal neurogenesis by lithium. J Neurochem 2000;75:1729–1734.
52 Chen RW, Chuang DM: Long-term lithium treatment suppresses p53 and Bax expression but increases Bcl-2 expression. A prominent role in neuroprotection against excitotoxicity. J Biol Chem 1999;274:6039–6042.
53 Lu R, Song L, Jope RS: Lithium attenuates p53 levels in human neuroblastoma SH-SY5Y cells. Neuroreport 1999;10:1123–1125.
54 Segal RA, Greenberg ME: Intracellular signaling pathways activated by neurotrophic factors. Annu Rev Neurosci 1996;19:463–489.
55 Tao X, Finkbeiner S, Arnold DB, Shaywitz AJ, Greenberg ME: Ca^{2+} influx regulates BDNF transcription by a CREB family transcription factor-dependent mechanism. Neuron 1998;20:709–726.

56 Corena-McLeod MDP, Oliveros A, Charlesworth C, Madden B, Liang YQ, Boules M, Shaw A, Williams K, Richelson E: Paliperidone as a mood stabilizer: a pre-frontal cortex synaptoneurosomal proteomics comparison with lithium and valproic acid after chronic treatment reveals similarities in protein expression. Brain Res 2008;1233:8–19.
57 Jovanovic JN, Czernik AJ, Fienberg AA, Greengard P, Sihra TS: Synapsins as mediators of BDNF-enhanced neurotransmitter release. Nat Neurosci 2000;3:323–329.
58 Rybakowski JK, Suwalska A, Skibinska M, Szczepankiewicz A, Leszczynska-Rodziewicz A, Permoda A, Czerski PM, Hauser J: Prophylactic lithium response and polymorphism of the brain-derived neurotrophic factor gene. Pharmacopsychiatry 2005;38:166–170.
59 Masui T, Hashimoto R, Kusumi I, Suzuki K, Tanaka T, Nakagawa S, Kunugi H, Koyama T: A possible association between the -116C/G single nucleotide polymorphism of the XBP1 gene and lithium prophylaxis in bipolar disorder. Int J Neuropsychopharmacol 2006;9:83–88.
60 Rybakowski JK, Suwalska A, Skibinska M, Dmitrzak-Weglarz M, Leszczynska-Rodziewicz A, Hauser J: Response to lithium prophylaxis: interaction between serotonin transporter and BDNF genes. Am J Med Genet B Neuropsychiatr Genet 2007;144B:820–823.
61 Dmitrzak-Weglarz M, Rybakowski JK, Suwalska A, Skibinska M, Leszczynska-Rodziewicz A, Szczepankiewicz A, Hauser J: Association studies of the BDNF and the NTRK2 gene polymorphisms with prophylactic lithium response in bipolar patients. Pharmacogenomics 2008;9:1595–1603.
62 Mamdani F, Alda M, Grof P, Young LT, Rouleau G, Turecki G: Lithium response and genetic variation in the CREB family of genes. Am J Med Genet B Neuropsychiatr Genet 2008;147B:500–504.
63 Riccio A, Ahn S, Davenport CM, Blendy JA, Ginty DD: Mediation by a CREB family transcription factor of NGF-dependent survival of sympathetic neurons. Science 1999;286:2358–2361.
64 Bonni A, Brunet A, West AE, Datta SR, Takasu MA, Greenberg ME: Cell survival promoted by the Ras-MAPK signaling pathway by transcription-dependent and -independent mechanisms. Science 1999;286:1358–1362.
65 Ruhe HG, Mason NS, Schene AH: Mood is indirectly related to serotonin, norepinephrine and dopamine levels in humans: a meta-analysis of monoamine depletion studies. Mol Psychiatry 2007;12:331–359.
66 Harro J, Oreland L: Depression as a spreading adjustment disorder of monoaminergic neurons: a case for primary implication of the locus coeruleus. Brain Res Brain Res Rev 2001;38:79–128.
67 Millan MJ: The role of monoamines in the actions of established and 'novel' antidepressant agents: a critical review. Eur J Pharmacol 2004;500:371–384.
68 Serretti A, Artioli P: From molecular biology to pharmacogenetics: a review of the literature on antidepressant treatment and suggestions of possible candidate genes. Psychopharmacology (Berl) 2004;174:490–503.
69 Serretti A, Lorenzi C, Lilli R, Smeraldi E: Serotonin receptor 2A, 2C, 1A genes and response to lithium prophylaxis in mood disorders. J Psychiatr Res 2000;34:89–98.
70 Serretti A, Lilli R, Lorenzi C, Franchini L, Di Bella D, Catalano M, Smeraldi E: Dopamine receptor D_2 and D_4 genes, GABA(A) alpha-1 subunit genes and response to lithium prophylaxis in mood disorders. Psychiatry Res 1999;87:7–19.
71 Serretti A, Lilli R, Lorenzi C, Franchini L, Smeraldi E: Dopamine receptor D_3 gene and response to lithium prophylaxis in mood disorders. Int J Neuropsychopharmcol 1998;1:125–129.
72 Serretti A, Lorenzi C, Lilli R, Mandelli L, Pirovano A, Smeraldi E: Pharmacogenetics of lithium prophylaxis in mood disorders: analysis of COMT, MAO-A, and Gbeta3 variants. Am J Med Genet 2002;114:370–379.
73 Dmitrzak-Weglarz M, Rybakowski JK, Suwalska A, Slopien A, Czerski PM, Leszczynska-Rodziewicz A, Hauser J: Association studies of 5-HT2A and 5-HT2C serotonin receptor gene polymorphisms with prophylactic lithium response in bipolar patients. Pharmacol Rep 2005;57:761–765.
74 Serretti A, Lilli R, Mandelli L, Lorenzi C, Smeraldi E: Serotonin transporter gene associated with lithium prophylaxis in mood disorders. Pharmacogenomics J 2001;1:71–77.
75 Serretti A, Malitas PN, Mandelli L, Lorenzi C, Ploia C, Alevizos B, Nikolaou C, Boufidou F, Christodoulou GN, Smeraldi E: Further evidence for a possible association between serotonin transporter gene and lithium prophylaxis in mood disorders. Pharmacogenomics J 2004;4:267–273.
76 Serretti A, Lilli R, Lorenzi C, Gasperini M, Smeraldi E: Tryptophan hydroxylase gene and response to lithium prophylaxis in mood disorders. J Psychiatr Res 1999;33:371–377.
77 Gray NA, Du J, Falke CS, Yuan P, Manji HK: Lithium regulates total and synaptic expression of the AMPA glutamate receptor GluR2 in vitro and in vivo. Ann NY Acad Sci 2003;1003:402–404.
78 Riederer P, Hoyer S: From benefit to damage. Glutamate and advanced glycation end products in Alzheimer brain. J Neural Transm 2006;113:1671–1677.
79 Kuiper MA, Teerlink T, Visser JJ, Bergmans PL, Scheltens P, Wolters EC: L-glutamate, L-arginine and L-citrulline levels in cerebrospinal fluid of Parkinson's disease, multiple system atrophy, and Alzheimer's disease patients. J Neural Transm 2000;107:183–189.
80 Sodhi M, Wood KH, Meador-Woodruff J: Role of glutamate in schizophrenia: integrating excitatory avenues of research. Expert Rev Neurother 2008;8:1389–1406.
81 Ferraro F, Ma XM, Sobota JA, Eipper BA, Mains RE: Kalirin/Trio Rho guanine nucleotide exchange factors regulate a novel step in secretory granule maturation. Mol Biol Cell 2007;18:4813–4825.
82 Shimshoni JA, Dalton EC, Jenkins A, Eyal S, Ewan K, Williams RS, Pessah N, Yagen B, Harwood AJ, Bialer M: The effects of central nervous system-active valproic acid constitutional isomers, cyclopropyl analogs, and amide derivatives on neuronal growth cone behavior. Mol Pharmacol 2007;71:884–892.
83 Williams RS, Cheng L, Mudge AW, Harwood AJ: A common mechanism of action for three mood-stabilizing drugs. Nature 2002;417:292–295.
84 Martinez A, Perez DI: GSK-3 inhibitors: a ray of hope for the treatment of Alzheimer's disease? J Alzheimers Dis 2008;15:181–191.
85 Wada A, Yokoo H, Yanagita T, Kobayashi H: Lithium: potential therapeutics against acute brain injuries and chronic neurodegenerative diseases. J Pharmacol Sci 2005;99:307–321.
86 Bauer M, Alda M, Priller J, Young LT: Implications of the neuroprotective effects of lithium for the treatment of bipolar and neurodegenerative disorders. Pharmacopsychiatry 2003;36(suppl 3):S250–S254.
87 Serretti A, Kato M, Kennedy JL: Pharmacogenetic studies in depression: a proposal for methodologic guidelines. Pharmacogenomics J 2008;8:90–100.
88 Mamdani F, Sequeira A, Alda M, Grof P, Rouleau G, Turecki G: No association between the PREP gene and lithium responsive bipolar disorder. BMC Psychiatry 2007;7:9.
89 Masui T, Hashimoto R, Kusumi I, Suzuki K, Tanaka T, Nakagawa S, Suzuki T, Iwata N, Ozaki N, Kato T, Kunugi H, Koyama T: Lithium response and Val66Met polymorphism of the brain-derived neurotrophic factor gene in Japanese patients with bipolar disorder. Psychiatr Genet 2006;16:49–50.
90 Rybakowski JK, Suwalska A, Czerski PM, Dmitrzak-Weglarz M, Leszczynska-Rodziewicz A, Hauser J: Prophylactic effect of lithium in bipolar affective illness may be related to serotonin transporter genotype. Pharmacol Rep 2005;57:124–127.
91 Washizuka S, Ikeda A, Kato N, Kato T: Possible relationship between mitochondrial DNA polymorphisms and lithium response in bipolar disorder. Int J Neuropsychopharmacol 2003;6:421–424.
92 Steen VM, Lovlie R, Osher Y, Belmaker RH, Berle JO, Gulbrandsen AK: The polymorphic inositol polyphosphate 1-phosphatase gene as a candidate for pharmacogenetic prediction of lithium-responsive manic-depressive illness. Pharmacogenetics 1998;8:259–268.

Neuropsychobiology

Neuropsychobiology 2010;62:72–78
DOI: 10.1159/000314708

Published online: May 8, 2010

The International Consortium on Lithium Genetics (ConLiGen): An Initiative by the NIMH and IGSLI to Study the Genetic Basis of Response to Lithium Treatment

Thomas G. Schulze[a, g, ε] Martin Alda[m, ε] Mazda Adli[h, ε] Nirmala Akula[a] Raffaella Ardau[t] Elise T. Bui[a] Caterina Chillotti[t] Sven Cichon[i, ζ] Piotr Czerski[v] Maria Del Zompo[t, u] Sevilla D. Detera-Wadleigh[a] Paul Grof[n, o, ε] Oliver Gruber[j] Ryota Hashimoto[x, δ] Joanna Hauser[v] Rebecca Hoban[b, c] Nakao Iwata[y, δ] Layla Kassem[a] Tadafumi Kato[z, δ] Sarah Kittel-Schneider[k] Sebastian Kliwicki[w] John R. Kelsoe[b, c] Ichiro Kusumi[β, δ] Gonzalo Laje[a] Susan G. Leckband[b, d, e] Mirko Manchia[u] Glenda MacQueen[p] Takuya Masui[β, δ] Norio Ozaki[γ, δ] Roy H. Perlis[f] Andrea Pfennig[l, ε] Paola Piccardi[u] Sara Richardson[a] Guy Rouleau[q] Andreas Reif[k] Janusz K. Rybakowski[w, ε] Johanna Sasse[l, ε] Johannes Schumacher[a, i] Giovanni Severino[u] Jordan W. Smoller[f] Alessio Squassina[u] Gustavo Turecki[r] L. Trevor Young[s, ε] Takeo Yoshikawa[α, δ] Michael Bauer[l, ε] Francis J. McMahon[a]

[a]Genetic Basis of Mood and Anxiety Disorders, National Institute of Mental Health, National Institutes of Health, Department of Health and Human Services, Bethesda, Md., [b]Department of Psychiatry, University of California San Diego, [c]Department of Psychiatry, VA San Diego Healthcare System, and [d]Department of Pharmacy, VA San Diego Healthcare System, La Jolla, Calif., [e]Skaggs School of Pharmacy and Pharmaceutical Sciences, University of California, San Diego, Calif., and [f]Department of Psychiatry, Massachusetts General Hospital and Harvard Medical School, Boston, Mass., USA; [g]Department of Genetic Epidemiology in Psychiatry, Central Institute of Mental Health, Mannheim, [h]Department of Psychiatry and Psychotherapy, Charité – Universitätsmedizin Berlin, Campus Charité Mitte, Berlin, [i]Department of Genomics, Life and Brain Center and Institute of Human Genetics, University of Bonn, Bonn, [j]Department of Psychiatry and Psychotherapy, Georg-August University, Göttingen, [k]Department of Psychiatry, Psychosomatics, and Psychotherapy, University of Würzburg, Würzburg, and [l]Department of Psychiatry and Psychotherapy, University Hospital Carl Gustav Carus, Technische Universität Dresden, Dresden, Germany; [m]Department of Psychiatry, Dalhousie University, Halifax, N.S., [n]Mood Disorders Center of Ottawa, Ottawa, Ont., [o]Department of Psychiatry, University of Toronto, Toronto, Ont., [p]Department of Psychiatry, University of Calgary, Calgary, Alta., [q]CHU Sainte-Justine Research Center, Department of Medicine, University of Montreal, and [r]Department of Psychiatry, Douglas Hospital Research Institute, McGill University, Montreal, Que., and [s]Department of Psychiatry, University of British Columbia, Vancouver, B.C., Canada; [t]Unit of Clinical Pharmacology, Hospital University Agency, and [u]Department of Neuroscience 'B.B. Brodie', University of Cagliari, Cagliari, Italy; [v]Psychiatric Genetic Unit, and [w]Department of Adult Psychiatry, Poznan University of Medical Sciences, Poznan, Poland; [x]Osaka University Graduate School of Medicine, Osaka, [y]Department of Psychiatry, Fujita Health University School of Medicine, Toyoake, [z]Laboratory for Molecular Dynamics of Mental Disorders, and [α]Laboratory for Molecular Psychiatry, RIKEN Brain Science Institute, Saitama, [β]Department of Psychiatry, Hokkaido University Graduate School of Medicine, Sapporo, [γ]Department of Psychiatry, Nagoya University Graduate School of Medicine, Nagoya, Japan; [δ]The Japanese Collaborative Group on the Genetics of Lithium Response in Bipolar Disorder; [ε]The International Group for the Study of Lithium-Treated Patients (IGSLI); [ζ]Institute of Neurosciences and Medicine (INM-1), Research Center Juelich, Juelich, Germany

KARGER

Fax +41 61 306 12 34
E-Mail karger@karger.ch
www.karger.com

Accessible online at:
www.karger.com/nps

Thomas G. Schulze, MD, Unit on the Genetic Basis of Mood and Anxiety Disorders
National Institute of Mental Health (NIMH), National Institutes of Health (NIH)
35 Convent Drive, Bldg. 35, Rm 1A205, MSC 3719
Bethesda, MD 20892-3719 (USA)
Tel. +1 301 451 7213, Fax +1 301 402 9081, E-Mail schulzet@mail.nih.gov

Key Words
Manic-depressive illness · Schizoaffective disorder · Mood stabilizer · Antidepressants · Suicidal behavior · Genome-wide association study · Neurogenesis · Neuroplasticity

Abstract
For more than half a decade, lithium has been successfully used to treat bipolar disorder. Worldwide, it is considered the first-line mood stabilizer. Apart from its proven antimanic and prophylactic effects, considerable evidence also suggests an antisuicidal effect in affective disorders. Lithium is also effectively used to augment antidepressant drugs in the treatment of refractory major depressive episodes and prevent relapses in recurrent unipolar depression. In contrast to many psychiatric drugs, lithium has outlasted various pharmacotherapeutic 'fashions', and remains an indispensable element in contemporary psychopharmacology. Nevertheless, data from pharmacogenetic studies of lithium are comparatively sparse, and these studies are generally characterized by small sample sizes and varying definitions of response. Here, we present an international effort to elucidate the genetic underpinnings of lithium response in bipolar disorder. Following an initiative by the International Group for the Study of Lithium-Treated Patients (www.IGSLI.org) and the Unit on the Genetic Basis of Mood and Anxiety Disorders at the National Institute of Mental Health, lithium researchers from around the world have formed the Consortium on Lithium Genetics (www.ConLiGen.org) to establish the largest sample to date for genome-wide studies of lithium response in bipolar disorder, currently comprising more than 1,200 patients characterized for response to lithium treatment. A stringent phenotype definition of response is one of the hallmarks of this collaboration. ConLiGen invites all lithium researchers to join its efforts.

Background

The articles in this special issue of *Neuropsychobiology* comprehensively review the use of lithium as a mood stabilizer in bipolar and unipolar affective disorders. They show that 60 years after Cade's discovery, lithium is still a first-line choice for prophylaxis in bipolar disorder. They furthermore discuss the evidence regarding lithium's antisuicidal effects, its use as an augmentation strategy in the treatment of unipolar depression, and provide novel insights into its neurobiological mechanisms of action. Finally, current pharmacogenetic knowledge about lithium treatment is reviewed. Taken together, however, these articles also highlight that, despite decades of lithium use in psychiatry and despite the current emphasis on the study of psychiatric genetics in modern biological psychiatry, pharmacogenetic data regarding lithium treatment have a tendency to be circumstantial and inconclusive.

Pharmacogenetics is a rapidly growing field that holds considerable promise for the development of medications that are more personalized and effective than those currently available. In all areas of medicine, pharmacogenetic studies of outcomes such as treatment response or characteristic side effects are on the rise; based on these findings, more and more pharmacogenetic tests are being offered and approved by the US Food and Drug Administration [1]. Pretreatment genetic testing has now even been added to the prescribing information for the anticoagulant warfarin [2]. Similarly, the Food and Drug Administration updated labeling for carbamazepine, recommending that patients of Asian ancestry be screened for the presence of the HLA allele B*1502 that has been implicated in carbamazepine-induced Stevens-Johnson syndrome and toxic epidermal necrolysis in Han Chinese people [3].

Sufficiently large, well-characterized samples as well as effective and efficient collaboration between academia and the pharmaceutical industry are among the critical prerequisites for success in the field of pharmacogenetics [4, 5]. Pharmacogenetic research in psychiatry has long been characterized by single lab efforts and small sample sizes. Only recently has our field witnessed large collaborative studies such as the Sequenced Treatment Alternatives to Relieve Depression (STAR*D) study (http://www.edc.pitt.edu/stard/) [6] in the United States, or the Genome-Based Therapeutic Drugs for Depression (GENDEP) project (http://gendep.iop.kcl.ac.uk) [7] in Europe, both of which study the pharmacogenetics of major depression. Indeed, the STAR*D and GENDEP projects have already generated several intriguing findings concerning the genetics of treatment response and side effects [8–14]. It is hoped that genome-wide association studies (GWAS) conducted in these and other samples will significantly increase our ability to guide the pharmacological treatment of psychiatric patients through the identification of genetic markers.

Notably, despite lithium's proven efficacy [15], to date there has been only one GWAS examining this 'pharmacological workhorse' of psychiatry [16]. In two cohorts encompassing more than 800 lithium-treated patients, multiple regions of interest were identified but none met the threshold for genome-wide significance. While intriguing, no adequately powered cohort yet exists to replicate and extend these findings. Here, we present a worldwide effort to address this situation: the international Consortium on Lithium Genetics (ConLiGen), spearheaded by researchers from the International Group for the Study of Lithium-Treated Patients (IGSLI) and the National Institute of Mental Health (NIMH).

The International Group for the Study of Lithium-Treated Patients

The IGSLI is an international group of scientists dedicated to lithium-related research, and its use in mental illness and mood disorders in particular. Founded in 1988 by Mogens Schou (Risskov/Aarhus, Denmark), Bruno Müller-Oerlinghausen (Berlin, Germany), and Paul Grof (Ottawa, Canada), the IGSLI has significantly contributed to lithium research over the past 20 years (www.igsli.org). Other scientists and centers have since joined the group, which currently comprises 35 members from Austria, Canada, the Czech Republic, Denmark, Germany, Poland, Switzerland, and the United States. The main goal of this group has been to conduct systematic work on those key questions regarding lithium treatment that can only be resolved by joint international effort. Unified designs have been created and scientific data from the IGSLI member centers have been linked for the purpose of shared analysis. This approach allows investigators to work with large numbers of prospectively followed patients – something that could only be accomplished via a multicenter approach. Overall, IGSLI research is based on shared, standardized, computer-based documentation of patients' diagnoses, family histories, course of illness before and during treatment, and on comparable modalities of treatment. The group meets regularly at research conferences to plan and discuss joint projects and to prepare publications.

At the 21st IGSLI meeting, which took place in late September 2007 in Dresden, Germany, the group discussed the results from the first, newly released GWAS of bipolar disorder, performed by researchers from the NIMH and Germany [17]. The strongest findings identified and replicated in this study were those encoding diacylglycerol kinase eta, a key protein in the lithium-sensitive phosphatidylinositol pathway and several genes in the *Wnt*-signaling cascade. Given the absence of a hypothesis-driven selection of single nucleotide polymorphisms in GWAS – a method more typical of candidate gene association studies – the observation that these findings implicated pathways relevant to lithium's mechanism of action was particularly intriguing. Spurred on by these findings, the IGSLI researchers concluded that studying these genes in samples that included data on patient response to lithium treatment could improve our understanding of how these genes determine response to lithium treatment and impact susceptibility to bipolar disorder. The IGSLI collaborators thus agreed to explore a framework that would allow researchers to engage in genetic studies of lithium response that were sufficiently powered. It was stated that such an endeavor should allow for participation by all bona fide lithium researchers within and beyond the IGSLI, while maintaining the highest possible level of stringency regarding phenotype definition.

May 6, 2008: The Consortium on Lithium Genetics Is Born

Following an invitation by IGSLI member Thomas G. Schulze and Francis J. McMahon, both from the NIMH's Unit on the Genetic Basis of Mood and Anxiety Disorders, prominent scientists in the field of lithium and bipolar genetic research met at the NIMH to discuss the possibility of creating an international consortium dedicated to the study of lithium pharmacogenetics. In attendance were (in alphabetical order): Martin Alda (Halifax, N.S., Canada), Michael Bauer (Dresden, Germany), Maria Del Zompo (via phone from Cagliari, Italy), Gonzalo Laje (Bethesda, Md., USA), Francis J. McMahon (Bethesda, Md., USA), Mirko Manchia (Cagliari, Italy), Roy H. Perlis (Boston, Mass., USA), Janusz K. Rybakowski (Poznan, Poland), Thomas G. Schulze (Bethesda, Md., USA), Johannes Schumacher (Bethesda, Md., USA), and Jordan W. Smoller (Boston, Mass., USA).

Reviewing evidence from the literature, and based on their own observations, the group emphasized the evident familiality in lithium treatment response, raising the possibility that genetic variation may contribute to interindividual differences in treatment response. If such differences could be identified, they might facilitate the development of novel treatments for bipolar disorder, or allow for better matching between patients and treatments. Over the last decade, the quest for a 'personalized medicine' approach in psychiatry has propelled a host of pharmacogenetic studies. Because of the lengthy trial-and-error process that currently characterizes the search for the most optimal treatment, pharmacogenetic studies in psychiatry have traditionally focused on treatment response or adverse effects associated with antidepressants or antipsychotic medications [18–22]. While initially limited by small sample sizes, pharmacogenetic studies in psychiatry have increasingly come to rely on large-scale collaborative efforts, such as STAR*D, GENDEP, or the Clinical Antipsychotic Trials of Intervention Effectiveness (CATIE) project. While some pharmacogenetic studies performed with these collaborative samples have produced intriguing results, difficulties in defining stringent target phenotypes across the various subsamples remain an important challenge [23, 24].

The researchers gathered at the NIMH on May 6, 2008 noted that, despite considerable and well-documented worldwide experience with lithium as an effective antimanic agent, mood stabilizer, and putative antisuicidal agent, there is a surprising dearth of large-scale pharmacogenetic studies of lithium treatment. We thus decided to create an international initiative whose goal would be to facilitate high-quality, well-powered analyses of lithium treatment response data that would ultimately allow for robust conclusions. The Consortium on Lithium Genetics, hereafter referred to as ConLiGen, was born.

ConLiGen's Scientific Goals

ConLiGen aims to identify genetic determinants of response to lithium treatment in bipolar disorder, as well as genetic determinants of adverse events emerging during lithium treatment. In the long run, ConLiGen may also study response to lithium treatment in general (e.g. lithium augmentation in the treatment of major depression).

Membership in ConLiGen

Any bona fide researcher or research group with access to samples of lithium-treated patients for whom DNA is available can join ConLiGen. Any new admission request is voted upon by ConLiGen members.

Communication between the ConLiGen Members

To ensure a constant exchange of ideas between members and allow for a straightforward realization of ConLiGen's goals, a monthly conference call is conducted. Furthermore, members meet once or twice a year at international meetings of various biological psychiatric organizations.

ConLiGen Advisory Board

An Advisory Board comprising international experts in the field of mood disorders research, and lithium research in particular, was established to offer ConLiGen an outside perspective as well as guidance on broad scientific directions, to serve as a liaison to nonacademic communities such as funding institutions, or industry, and finally, to act as one of ConLiGen's publicly visible faces. Currently, the following researchers are members of the Advisory Board (in alphabetical order): Robert H. Belmaker (Division of Psychiatry, Ben Gurion University of the Negev, Beersheva, Israel), Gian Luigi Gessa (Department of Neuroscience 'B.B. Brodie', University of Cagliari, Cagliari, Italy), Paul Greengard (Laboratory of Molecular and Cellular Neuroscience, Rockefeller University, New York, N.Y., USA), Kay R. Jamison (Department of Psychiatry and Behavioral Sciences, Johns Hopkins University School of Medicine, Baltimore, Md., USA), Richard S. Jope (Department of Psychiatry and Behavioral Neurobiology, University of Alabama at Birmingham, Birmingham, Ala., USA), Husseini K. Manji (CNS & Pain, Johnson and Johnson Pharmaceutical Research and Development, Titusville, N.J., USA), and Leon E. Rosenberg (Department of Molecular Biology and the Woodrow Wilson School of Public and International Affairs, Princeton University, Princeton, N.J., USA).

Phenotype Definition of Lithium Response: A Major Prerequisite for Pharmacogenetic Studies of Lithium

ConLiGen's first and most crucial goal is to define the phenotype of lithium response. Treatment response is a complex construct that requires researchers to make judgments about adequacy of treatment and tolerability as well as assess changes in episode frequency or symptom severity. In many cases this information must be assessed retrospectively, with the inherent limitations associated with recall bias, missing information, or the fact that the treatment has not followed a strict research protocol. One scale that incorporates such data is an 11-point scale developed by Martin Alda and colleagues [25]

Name: ______________________ Date: ______________

Criterion A

The criterion A is used to determine an association between clinical improvement and the treatment. The rating should apply to the period of treatment considered adequate in duration and dosage. The illness activity should be judged by frequency, severity, and duration of episodes.

10 = Complete response, no recurrences in the course of adequate treatment, no residual symptoms, and full functional recovery
9 = Very good response, no recurrences, but the patient may have minimal residual symptoms (transient anxiety, sleep disturbance, dysphoria, irritability) not requiring any intervention
8 = Very good response. Illness activity reduced by more than 90%
7 = Good response. Illness activity reduced by 80 - 90 %
6 = Good response. Reduction in activity of illness by 65 - 80%
5 = Moderate response. Reduction in illness activity by 50 - 65%
4 = Moderate improvement. Reduction in illness activity by 35 - 50%
3 = Mild improvement. Reduction of illness activity by 20 - 35%
2 = Mild improvement. Reduction of illness activity by 10 - 20%
1 = Minimal improvement. Reduction of illness activity by 0 - 10%
0 = No change or worsening

A Criterion Score: _______

Criteria B

The criteria B are used to establish whether there is a causal relationship between clinical improvement and the treatment. Score 0, 1 or 2 points for each item:

B1: Number of episodes off the treatment.

0 = 4 or more episodes
1 = 2 or 3 episodes
2 = 1 episode

B1: _____

Drug: ___________ **Evaluated By:** __________

B2: Frequency of episodes off the treatment.

0 = Average to high, including rapid cycling
1 = Low, spontaneous remissions of 3 or more years on average
2 = 1 episode only, risk of recurrence cannot be established

B2: _____

B3: Duration of the treatment.

0 = 2 or more years
1 = 1 - 2 years
2 = Less than 1 year

B3: _____

B4: Compliance during period(s) of stability.

0 = Excellent, e.g. documented by drug levels in the therapeutic range
1 = Good, more than 80% levels in the therapeutic range
2 = Poor, repeatedly off treatment, less than 80% levels in the therapeutic range

B4: _____

B5: Use of additional medication during the period of stability.

0 = None except infrequent sleep medication (1 per week or less); no other mood stabilizers, antidepressants or antipsychotics for control of mood symptoms
1 = Low-dose antidepressants or antipsychotics as an "insurance", or prolonged use of sleep medication
2 = Prolonged or systematic use of an antidepressant or antipsychotic

B5: _____

B Criteria Score: _____

Total Scale Score: _____
(Subtract B from A)

Fig. 1. Retrospective criteria of long-term treatment response in research subjects with bipolar disorder.

(fig. 1); other approaches include longitudinal outcome measures that consider time to recurrence or symptom burden during treatment [16, 26].

The 11-point scale measures the extent of improvement during long-term treatment. The scale's A score is a composite measure of change in frequency, duration, and severity of illness episodes in the course of lithium treatment. It is weighted by factors that influence the degree to which the observed clinical change is considered to be due to lithium (B1–B5 scores in the scale). The scale has been developed in the context of a study assessing response to treatment in subjects not followed according to a research protocol, namely relatives of probands in our genetic studies [25]. Subsequently, it has been widely used in several other studies at IGSLI centers [27–29] and at other centers involved in lithium research [pers. commun. from John Kelsoe, San Diego, Calif., USA and Maria del Zompo, Cagliari, Italy], which imparts face validity. Within ConLiGen, phenotypic assessment will be based on any available information including life charts when available and quantified using the scale; interrater reliability meetings will be organized, facilitated by ConLiGen member Martin Alda, and case vignettes will also be reviewed to establish between-center reliability.

Variables describing treatment tolerability or side effects may be studied in subsequent projects. Because the issue of 'best response phenotype' is far from trivial, ConLiGen will strive to continuously weigh evidence from future clinical and biological studies of lithium in an effort to refine the definition of phenotype response. Evaluating response to long-term treatment in an illness with a highly variable natural course presents a challenge. Many patients with bipolar disorder experience spontaneous remissions of variable timing and duration. Moreover, in a pharmacogenetic study we need to evaluate the quality of response not for groups of subjects as in clinical trials but individually for each patient. While prospective studies will be able to implement more precise measures, our approach is a practical way to assess the quality of response in a variety of patients treated in diverse settings.

ConLiGen's Current Project and Long-Term Mission

ConLiGen is poised to assess all aspects of the pharmacogenetics of lithium treatment in psychiatric disorders, including the study of genetic susceptibility to potential treatment-emergent adverse events (e.g. weight gain, hypothyroidism, tremor). As its first project, ConLiGen intends to conduct a GWAS of stringently defined response to lithium treatment in bipolar disorder. ConLiGen members and the various research centers which they are affiliated with are joining their samples for a centralized genotyping effort to be performed at the Unit on the Genetic Basis of Mood and Anxiety Disorders of the NIMH and the Department of Genomics of the Life and Brain Center at the University of Bonn, Germany. For the primary projects, a previously validated scale will be used to define response to lithium treatment, as described above. Individuals scoring between 7 and 10 will be considered lithium 'responders', while individuals with scores between 0 and 6 will be considered 'nonresponders'. Presently, the total sample comprises more than 1,200 bipolar patients for whom response to lithium treatment has been or is currently being assessed by means of the scale. From preliminary analyses conducted in select IGSLI samples (data not shown), we can assume that about 35–40% of patients will qualify as responders. Previous studies [8, 9] suggest larger genetic effect sizes (e.g. allelic odds ratios between 1.5 and 2) for a narrowly defined pharmacogenetic phenotype than for a categorically defined clinical diagnosis. Thus, assuming a minor allele frequency of 0.3 and genotype relative risks of 1.4 for individuals heterozygous, and of 1.96 for individuals homozygous for the risk allele, the combined ConLiGen sample will have a power of 83% to detect an effect at a significance level of 1×10^{-8} [30].

Although the combined ConLiGen sample will be the largest sample to date to investigate lithium response on a genome-wide scale, we are aware that any finding, regardless of whether it reaches levels of genome-wide significance, will ultimately have to be confirmed in independent samples. Thus, ConLiGen's mission will not be finished after the completion of its GWAS. On the contrary, ConLiGen will continue to invite researchers to join its efforts in order to increase the available sample size of patients adequately characterized for lithium response. In collaboration with both IGSLI centers and large, long-standing multicenter projects such as the NIMH Bipolar Disorder Genetics Initiative, ConLiGen will be actively engaged in supporting and organizing urgently needed prospective studies of lithium response in bipolar disorder and other conditions.

Since Cade discovered lithium's beneficial effects in the treatment of bipolar disorder 60 years ago, this agent has become almost synonymous with the treatment of bipolar disorder worldwide [15]. Yet, little is known about the genetic underpinnings of lithium response or the development of side effects associated with its use. In a scientific environment characterized by calls for personalized medicine and the growth of large-scale pharmacogenetic studies in many fields of medicine, ConLiGen's goal is to put lithium at the forefront of pharmacogenetic studies in psychiatry.

Acknowledgments

This research was in part supported by the Intramural Program of the NIMH. Dr. Thomas G. Schulze's work is furthermore supported by NARSAD. Dr. Martin Alda's work on the phenotype of lithium response has been supported by grants from the Canadian Institutes of Health Research (CIHR, grant 64410) and by NARSAD. Dr. Smoller's work is supported in part by a grant from the NIMH (MH079799). Dr. Reif is supported by the Deutsche Forschungsgemeinschaft (DFG) (KFO 125 and SFB TRR 58) and the German Ministry for Education and Research (BMBF) (IZKF N-4). The work of the Cagliari group was partly supported by the Regional Councillorship of Health, 'Regione Autonoma della Sardegna'.

References

1 Swen JJ, Huizinga TW, Gelderblom H, de Vries EG, Assendelft WJ, Kirchheiner J, Guchelaar HJ: Translating pharmacogenomics: challenges on the road to the clinic. PLoS Med 2007;4:e209.

2 Wadelius M, Sörlin K, Wallerman O, Karlsson J, Yue QY, Magnusson PK, Wadelius C, Melhus H: Warfarin sensitivity related to CYP2C9, CYP3A5, ABCB1 (MDR1) and other factors. Pharmacogenomics J 2004;4:40–48.

3 Chung WH, Hung SI, Hong HS, Hsih MS, Yang LC, Ho HC, Wu JY, Chen YT: Medical genetics: a marker for Stevens-Johnson syndrome. Nature 2004;428:486.

4 Penny MA, McHale D: Pharmacogenomics and the drug discovery pipeline: when should it be implemented? Am J Pharmacogenomics 2005;5:53–62.

5 Roses AD: Pharmacogenetics in drug discovery and development: a translational perspective. Nat Rev Drug Discov 2008;7:807–817.

6 Fava M, Rush AJ, Trivedi MH, Nierenberg AA, Thase ME, Sackeim HA, Quitkin FM, Wisniewski S, Lavori PW, Rosenbaum JF, Kupfer DJ: Background and rationale for the sequenced treatment alternatives to relieve depression (STAR*D) study. Psychiatr Clin North Am 2003;26:457–494.

7 Uher R, Farmer A, Maier W, Rietschel M, Hauser J, Marusic A, Mors O, Elkin A, Williamson RJ, Schmael C, Henigsberg N, Perez J, Mendlewicz J, Janzing JG, Zobel A, Skibinska M, Kozel D, Stamp AS, Bajs M, Placentino A, Barreto M, McGuffin P, Aitchison KJ: Measuring depression: comparison and integration of three scales in the GENDEP study. Psychol Med 2008;38:289–300.

8 McMahon FJ, Buervenich S, Charney D, Lipsky R, Rush AJ, Wilson AF, Sorant AJ, Papanicolaou GJ, Laje G, Fava M, Trivedi MH, Wisniewski SR, Manji H: Variation in the gene encoding the serotonin 2A receptor is associated with outcome of antidepressant treatment. Am J Hum Genet 2006;78:804–814.

9 Laje G, Paddock S, Manji H, Rush AJ, Wilson AF, Charney D, McMahon FJ: Genetic markers of suicidal ideation emerging during citalopram treatment of major depression. Am J Psychiatry 2007;164:1530–1538.

10 Perlis RH, Purcell S, Fava M, Fagerness J, Rush AJ, Trivedi MH, Smoller JW: Association between treatment-emergent suicidal ideation with citalopram and polymorphisms near cyclic adenosine monophosphate response element binding protein in the STAR*D study. Arch Gen Psychiatry 2007;64:689–697.

11 Paddock S, Laje G, Charney D, Rush AJ, Wilson AF, Sorant AJ, Lipsky R, Wisniewski SR, Manji H, McMahon FJ: Association of GRIK4 with outcome of antidepressant treatment in the STAR*D cohort. Am J Psychiatry 2007;164:1181–1188.

12 Perlis RH, Moorjani P, Fagerness J, Purcell S, Trivedi MH, Fava M, Rush AJ, Smoller JW: Pharmacogenetic analysis of genes implicated in rodent models of antidepressant response: association of TREK1 and treatment resistance in the STAR(*)D study. Neuropsychopharmacology 2008;33:2810–2819.

13 Lekman M, Laje G, Charney D, Rush AJ, Wilson AF, Sorant AJ, Lipsky R, Wisniewski SR, Manji H, McMahon FJ, Paddock S: The FKBP5-gene in depression and treatment response – an association study in the Sequenced Treatment Alternatives to Relieve Depression (STAR*D) Cohort. Biol Psychiatry 2008;63:1103–1110.

14 Uher R, Huezo-Diaz P, Perroud N, Smith R, Rietschel M, Mors O, Hauser J, Maier W, Kozel D, Henigsberg N, Barreto M, Placentino A, Dernovsek MZ, Schulze TG, Kalember P, Zobel A, Czerski PM, Larsen ER, Sourey D, Giovannini C, Gray JM, Lewis CM, Farmer A, Aitchison KJ, McGuffin P, Craig I: Genetic predictors of response to antidepressants in the GENDEP project. Pharmacogenomics J 2009;9:225–233.

15 Bauer M, Grof P, Müller-Oerlinghausen B (eds): Lithium in Neuropsychiatry: The Comprehensive Guide. London, Informa Healthcare, 2006.

16 Perlis RH, Smoller JW, Ferreira MAR, McQuillin A, Bass N, Chir MB, Lawrence J, Sachs GS, Nimgaonkar V, Scolnick EM, Gurling H, Sklar P, Purcell S: A genomewide association study of response to lithium for prevention of recurrence in bipolar disorder. Am J Psychiatry 2009;166:718–725.

17 Baum AE, Akula N, Cabanero M, Cardona I, Corona W, Klemens B, Schulze TG, Cichon S, Rietschel M, Nöthen MM, Georgi A, Schumacher J, Schwarz M, Abou Jamra R, Höfels S, Propping P, Satagopan J, Detera-Wadleigh SD, Hardy J, McMahon FJ: A genome-wide association study implicates diacylglycerol kinase eta (DGKH) and several other genes in the etiology of bipolar disorder. Mol Psychiatry 2008;13:197–207.

18 Rietschel M, Kennedy JL, Macciardi F, Meltzer HY: Application of pharmacogenetics to psychotic disorders: the first consensus conference. The Consensus Group for Outcome Measures in Psychoses for Pharmacological Studies. Schizophr Res 1999;37:191–196.

19 Lerer B, Segman RH, Fangerau H, Daly AK, Basile VS, Cavallaro R, Aschauer HN, McCreadie RG, Ohlraun S, Ferrier N, Masellis M, Verga M, Scharfetter J, Rietschel M, Lovlie R, Levy UH, Meltzer HY, Kennedy JL, Steen VM, Macciardi F: Pharmacogenetics of tardive dyskinesia: combined analysis of 780 patients supports association with dopamine D_3 receptor gene Ser9Gly polymorphism. Neuropsychopharmacology 2002;27:105–119.

20 Malhotra AK, Murphy GM Jr, Kennedy JL: Pharmacogenetics of psychotropic drug response. Am J Psychiatry 2004;161:780–796.

21 Correll CU, Malhotra AK: Pharmacogenetics of antipsychotic-induced weight gain. Psychopharmacology (Berl) 2004:477–489.

22 Malhotra AK, Lencz T, Correll CU, Kane JM: Genomics and the future of pharmacotherapy in psychiatry. Int Rev Psychiatry 2007;19:523–530.

23 van den Oord EJ, Adkins DE, McClay J, Lieberman J, Sullivan PF: A systematic method for estimating individual responses to treatment with antipsychotics in CATIE. Schizophr Res 2009;107:13–21.

24 Need AC, Keefe RS, Ge D, Grossman I, Dickson S, McEvoy JP, Goldstein DB: Pharmacogenetics of antipsychotic response in the CATIE trial: a candidate gene analysis. Eur J Hum Genet 2009;17:946–957.

25 Grof P, Duffy A, Cavazzoni P, Grof E, Garnham J, MacDougall M, O'Donovan C, Alda M: Is response to prophylactic lithium a familial trait? J Clin Psychiatry 2002;63:942–947.

26 Perlis RH, Ostacher MJ, Patel JK, Marangell LB, Zhang H, Wisniewski SR, Ketter TA, Miklowitz DJ, Otto MW, Gyulai L, Reilly-Harrington NA, Nierenberg AA, Sachs GS, Thase ME: Predictors of recurrence in bipolar disorder: primary outcomes from the Systematic Treatment Enhancement Program for Bipolar Disorder (STEP-BD). Am J Psychiatry 2006;163:217–224.

27 Garnham J, Munro A, Slaney C, Macdougall M, Passmore M, Duffy A, O'Donovan C, Teehan A, Alda M: Prophylactic treatment response in bipolar disorder: results of a naturalistic observation study. J Affect Disord 2007;104:185–190.

28 Passmore MJ, Garnham J, Duffy A, MacDougall M, Munro A, Slaney C, Teehan A, Alda M: Phenotypic spectra of bipolar disorder in responders to lithium versus lamotrigine. Bipolar Disord 2003;5:110–114.

29 Hajek T, Hahn M, Slaney C, Garnham J, Green J, Ruzickova M, Zvolsky P, Alda M: Rapid cycling bipolar disorders in primary and tertiary-care treated patients. Bipolar Disord 2008;10:495–502.

30 Purcell S, Cherny SS, Sham PC: Genetic Power Calculator: design of linkage and association genetic mapping studies of complex traits. Bioinformatics 2003;19:149–150.

Neuropsychobiology

Author Index Vol. 62, No. 1, 2010

KARGER
Fax +41 61 306 12 34
E-Mail karger@karger.ch
www.karger.com

Accessible online at:
www.karger.com/nps

Neuropsychobiology

Subject Index Vol. 62, No. 1, 2010

KARGER

Fax +41 61 306 12 34
E-Mail karger@karger.ch
www.karger.com

Accessible online at:
www.karger.com/nps

F10532

Major pharmacogenomic findings in psychiatric medicine

Pharmacogenomics in Psychiatry

Editors
Matthias Schwab
Wolfgang P. Kaschka
Edoardo Spina

Psychiatry; Pharmacology, Genetics

Contents

In recent years, there have been major developments in the fields of pharmacogenetics and pharmacogenomics, with the potential to make drug treatments in psychiatric medicine more effective. However, improvements in drug efficacy and tolerability, as well as finding the optimal dosage, can only be realized if in vivo mechanisms of drug action and ADME (absorption, distribution, metabolism, excretion) processes (pharmacokinetics) of psychopharmacological agents are better understood.
In this volume, current progress and perspectives in pharmacogenetic testing of drug-metabolizing enzymes, drug transporters and other drug targets involved in the response to psychotropic agents are described extensively. This provides a timely overview of what has been achieved in the area of psychiatric pharmacogenomics alongside some promising directions and perspectives for future research.
Psychiatrists, general medical doctors as well as pharmacologists and clinical pharmacologists will find new insights into the development and applications of pharmacogenomics in psychiatry.

www.karger.com/adbip

KI10439

Advances in Biological Psychiatry, Vol. 25 →
Series Editors: Ebert, D. (Freiburg); Ebmeier, K.P. (Oxford); Gattaz, W.F. (São Paulo); Kaschka, W.P. (Ulm/Ravensburg)
ISSN 0378–7354 / e-ISSN 1662–2274

Pharmacogenomics in Psychiatry
Editors: Schwab, M. (Stuttgart/Tübingen); Kaschka, W.P. (Ulm/Ravensburg); Spina, E. (Messina)
VIII + 140 p., 8 fig., 4 in color, 17 tab., hard cover, 2010
CHF 64.– / EUR 45.50 / USD 64.00
Prices subject to change
EUR price for Germany, USD price for USA only
ISBN 978–3–8055–9498–1
e-ISBN 978–3–8055–9499–8

KARGER

Order Form

Please send: ____ copy/ies

Postage and handling free with prepayment

Payment:
Please charge to my credit card
☐ American Express ☐ Diners ☐ Eurocard
☐ MasterCard ☐ Visa

Card No.: ____________

Exp. date: ____________

CVV/CVC ____________
(3 digits in the signature field on the back of Visa and MasterCard)

☐ Check enclosed ☐ Please bill me

Orders may be placed with any bookshop, subscription agency, directly with the publisher or through a Karger distributor.

Fax: +41 61 306 12 34

S. Karger AG, P.O. Box, CH–4009 Basel (Switzerland)
E-Mail orders@karger.ch, **www.karger.com**

Name/Address: ____________

____

____

____

____

Date: ____________

Signature: ____________